Structures of Scientific Collaboration

Inside Technology
edited by Wiebe E. Bijker, W. Bernard Carlson, and Trevor Pinch

For a list of the series, see page 275.

Structures of Scientific Collaboration

Wesley Shrum
Joel Genuth
Ivan Chompalov

The MIT Press
Cambridge, Massachusetts
London, England

© 2007 Massachusetts Institute of Technology

All rights reserved. No part of this book may be reproduced in any form by any electronic or mechanical means (including photocopying, recording, or information storage and retrieval) without permission in writing from the publisher.

For information on quantity discounts, email special_sales@mitpress.mit.edu.

Set in Baskerville by The MIT Press. Printed and bound in the United States of America.

Library of Congress Cataloging-in-Publication Data

Structures of scientific collaboration / edited by Wesley Shrum, Joel Genuth, and Ivan Chompalov
  p. cm. — (Inside technology)
Includes bibliographical references and index.
ISBN-13: 978-0-262-19559-1 (hardcover : alk. paper)
1. Research—International cooperation. 2. Physics—Research—International cooperation. 3. Intellectual cooperation. 4. Academic-industrial collaboration.
I. Shrum, Wesley, 1953– II. Carlson, W. Bernard. III. Chompalov, Ivan.
Q172.5.I5S77 2007
530.072'4—dc22

2006033370

10 9 8 7 6 5 4 3 2 1

Contents

Preface   vii

Introduction   *1*

1
Formation   *25*

2
Magnitude   *67*

3
Organization   *87*

4
Technology and Interdependence   *119*

5
Trust, Conflict, and Performance   *151*

Conclusion   *195*

Appendix A: Organizational Approaches to Collaboration   *217*
Appendix B: Selected Case Studies   *223*
Notes   *231*
References   *259*
Series List   *275*
Index   *277*

Preface

Strolling down Bourbon Street in New Orleans, three authors were discussing *On Any Sunday*, a football film by Oliver Stone. The first described a scene in which an experienced coach tries to explain to his star quarterback what players miss after retirement. "It's not the winning or the awards. It's not the fans or the cheering or the locker room. What you miss is that feeling of eleven people, all facing the same direction, all moving down the field together." The second, not much of an athlete himself, lamented that he would never have that feeling. "What do you mean?" said the third. "Aren't the three of us all here, right now, moving down the field together?"

He was right, of course. More important, he was right in a way that highlights the similarities and differences between collaboration and cooperation, two concepts that are often confused. A football team is one kind of collaboration. Players in fixed structural positions seek to execute a play, kick a field goal, prevent a first down. Their roles are limited and specific, working together to outscore an opponent. Since they are human beings, they like it when they win, but they like it even better when they actively help each other along the way—a process that may also be called teamwork. Three friends exploring a city together have determined, for the present, to *take each other's interest into account*. There is no specific "objective" for their cooperative interactions, except in the expressive sense of enjoying each other's company. Such important but diffuse activities are inherent in all social interactions with positive affect, and they are distinct from instrumental goals such as manufacturing a car, building a church, or writing a book.

Taking the interests of others into account is the basis of cooperation, a fundamental social process. Collaboration—our main interest here—involves something more, a shared objective. The shared objective is what brings social actors together and induces them to create plans, procedures, and lasting structures. It is difficult to imagine collaboration without cooperation among individuals in some minimal sense. But collaboration itself is not a form of social interaction on the same plane as cooperation, conflict, and competition: a goal, or a set of goals, is crucial. Considered as a social *object*, collaboration is an aggregation of social actors or identities. Often it is intendedly ephemeral, in which case it is distinct from complex organizations that are designed to last. When goals are met or resources disappear, collaborations expire. Considered as a social *process*, collaboration is a way of organizing, oriented toward common objectives. One implication of this characterization of collaboration versus cooperation is that organizations are capable of only the former, while people can do both. Only human beings can take each other's interests into account.

For individuals to collaborate effectively, individual interests must sometimes be sacrificed for the sake of the common objective. The threats and rewards of working within an organization have traditionally supplemented moral commitment to ensure that such objectives are met. Today, collaboration among organizations is a growing phenomenon in all spheres of human activity, science and technology included. Modernization, globalization, new communication technologies, and competition in the efficient utilization of human, financial, and technological resources have brought about a general tendency to develop knowledge in new organizational structures. In a variety of fields, it is rapidly becoming the norm for researchers from different locations and institutions to work together on common projects: scientific and technological collaborations are part of a general trend toward more fluid, flexible, and temporary organizational arrangements. Joint ventures, strategic alliances, consortia, partnerships, and all manner of networks characterize many sectors of the economy. As modern production and services become increasingly knowledge-based, and as technological innovation becomes more complex and diversified, organizations begin more and more to interpenetrate—entities combine and recombine in processes that blur traditional enterprise boundaries. We assume all this as background. Our starting point is that science and technology are integral to the networking dynamic that characterizes the modern social order, and that scientific work itself has been affected by these global tendencies that involve collaboration.

Of course, our work is collaborative. More important, we did not initiate the project that gave rise to our work. The origin of this study dates to the late 1970s, when historians and archivists at the American

Institute of Physics' Center for the History of Physics conducted a study of U.S. national laboratories and recognized that no one was attempting to document the activities of multi-institutional collaborations of researchers who were conducting experiments at the national laboratories. The collaborations were disbanding after completing their experiments without any assessment of the records they had created and without any provisions for preserving their historically or scientifically valuable records.

In 1989, AIP archivist Joan Warnow-Blewett initiated a long-term study to address this problem of documentation. She enlisted historians and sociologists to develop the knowledge that archivists and administrators would need to identify records that should be preserved for posterity. She sought and received funding from the National Science Foundation, the Department of Energy, the National Historic Preservation and Records Commission, and the Mellon Foundation. (The National Science Foundation, under its Program in Science and Technology Studies, later supported our integration of the three phases of the AIP study into one data set.) The first phase examined 24 collaborations in high-energy physics. The second examined six collaborations in space science and eight in geophysics. The third phase examined 23 collaborations in ground-based astronomy, materials science, heavy-ion and nuclear physics, medical physics, and computer-mediated collaborations.

In all, about 600 participants in these collaborations were interviewed about their origins, their organization, their management, their communication practices, the relationships among scientists, and the collaborations' outcomes. The objective was to build an empirical base for understanding how records have been created, used, and collected. The current volume is based on historical and sociological examination of these interviews, focusing on the structure of scientific collaborations. After all the interviews had been transcribed and all the reports written, we set out to reanalyze and interpret the whole. It is good to write a small book on a large project.

Joan Warnow-Blewett died in May 2006. We dedicate this volume to her memory. Her energy and zeal were obvious to her most casual acquaintances. Her passion for documentation research was legendary among her colleagues. She was the Principal Investigator of the AIP Study of Multi-Institutional Collaborations in Physics and Allied Sciences. There is no way to overstate our gratitude and indebtedness to her. R. Joseph Anderson, her successor, helped in more ways than we can count, even before taking over as archivist. He sat with us for hours, argued with us about our methodology, and greatly improved our study.

We honor Spencer Weart, director of the AIP Center for History of Physics, as a constant source of constructive criticism. He was convinced that this book should be written, he encouraged us to write it, and he improved it through his comments. We thank the AIP staff and consultants who conducted interviews with us over the years. A study such as this cannot be conducted without considerable expertise in the subject matter, and we consider ourselves extremely lucky to have worked with Anthony Capitos, Ron Doel, John Krige, Janet Linde, Lynn Maloney, Frederick Nebeker, and Arturo Russo. The late David Edge and Mike Lynch, successive editors of *Social Studies of Science*, were instrumental in helping us refine our notions of trust and performance. We are very much indebted to Edward Hackett and Ulrike Felt for assistance and feedback during several phases of the project, as well as to our friend and editor Wiebe Bijker, who helped us immensely in bringing the volume to fruition. Near or far, one cannot wish for better colleagues than Ed, Uli, and Wiebe.

Since one may also be indebted to an entire community, we are grateful for the annual meetings of the Society for Social Studies of Science and the History of Science Society, which have been receptive venues over the years. In particular, the intellectual development of our ideas owes a great deal to Karin Knorr Cetina, Michel Callon, Sheila Jasanoff, and Bruno Latour. Many scientists, historians, and scholars provided special help on certain phases of the project. For high-energy physics our main consultants were Peter Galison, Roxanne Nilan, Lynne Zucker, and Lowell Hargens. Others who provided advice were Helen Samuels, Anne Millbrooke, Sharon Gibbs Thibodeau, Richard McKay, Victoria Davis, Louise Addis, David Berley, Sheldon Glashow, Bernard Hildebrand, Herbert Kinney, William Kirk, Ronald Kline, Derek Lowenstein, Robert Smith, William Wallenmeyer, Robert Wilson, Stanley Wojcicki, and Harriet Zuckerman. For space science and geophysics we also relied heavily on Naomi Oreskes and Deborah Day. Additional advice and criticism came from Arthur Davidsen, Stewart Gillmor, Robert Heinmiller, Frank McDonald, Chandra Mukerji, John Naugle, William Nierenberg, Arthur Nowell, Kenneth Pedersen, John Perry, Charles Prewitt, Mary Rabinowitch, Jeffrey Rosendhal, and James Van Allen. For the last phase, we also consulted Gabriel Aeppli, Alan Berson, Robert Dickman, Paul Fleury, Phillip Geil, Russell Huebner, Thomas Humanic, Goetz Oertel, Thomas Rosenbaum, Ulrich Strom, Paul Vanden Bout, John Watson, Sidney Wolff, John Knauss, Robert Williams, Joseph

Alexander, William Maher, Elizabeth Adkins, Mary Ann Wallace, Marie Allen, Larry Baume, and Terry Cook.

Collaboration is impossible without family and friends. Among those who worked with us (and, fortunately, without us) were Susan K. Arnold, Steve Coffee, D. Richard Johnson, Mike Rood, George Clark, Steve Allison, Kenneth Kimmel, John Harms, Gary Slavik, Scott Sidney, Roger Barnes, Wesley Monroe Shrum (the real one), Mike Grimes, Bill Bankston, Jeanne Hurlbert, Jack Beggs, Rick Weil, Ed Shihadeh, Frank Taylor, Susan Whitcome, Peter Chompalov, Abigail and Ken Cooper, Saul and Molly Genuth, Katherine Hayes and Tom Moore, Jonathan and Barbara Polonsky, and especially Miriam and Naomi Genuth.

We thank the publisher of *Research Policy*, the publisher of *Minerva*, and the publisher of *Social Studies of Science* for permission to reprint passages and tables.

We are eternally indebted to our informants, the scientists, engineers, and managers who described with wit and detail some of the most fascinating collaborations in twentieth-century science. Although we have removed their names from the quotations in this volume, almost all generously allowed their interview transcripts to be archived at the American Institute of Physics and made available for scholarly research at the AIP's Niels Bohr Library. We hope to persuade our readers that the transcripts are well worth using.

Structures of Scientific Collaboration

Introduction

The collaborating scientists, both Japanese and American, were ecstatic. Their historic discovery of neutrino mass was about to be announced at an international meeting to an audience of 300 physicists. The demonstration, convincing beyond a reasonable doubt to those present, showed that neutrinos, created from the impact of cosmic rays on the upper atmosphere, change form as they pass through the Earth. Although the mass of the neutrino itself had not been measured, neutrinos could now be held to account for a significant portion of the mass in the universe.

But even as participants argued about the number of Nobel Prizes that would result, a pall fell over the conference. The day after the discovery, major cuts in funding for the Kamioka Neutrino Observatory were announced by the Japanese government. Owing to budget shortfalls, an immediate 15 percent reduction would be followed by another 15 percent the following year. The Super-Kamiokande neutrino detector, costing $100 million to build, would have to be shut down for an extended period.[1]

"Big Science"

The aforementioned episode in the recent history of science illustrates both the intimate connection of modern science with political decision making and a blind spot in the way we interpret that connection. The Kamioka project is plainly "Big Science" in the sense originally intended by Derek Price and Alvin Weinberg. Kamioka absorbs the professional efforts of 120 scientists, and its large scale requires significant government funding. These resource requirements brought the project within the purview of Japanese national politics, and its decision makers chose to cut funding to cope with budgetary difficulties. The result is a morality tale for our times: Virtuous scientists with dreams of enlightening humanity, beware the power of bureaucrats who live in fear of unbalanced books, irregular procedures, and low approval ratings—they will cut you off even when Nobel-caliber results are within reach.

We, however, find the story of Kamioka telling for reasons that involve the composition and the organization of large scientific collaborations. The prospect of scientific discovery requires significant technological apparatus—in this case a ten-story water tank built under a mountain range—that serves as a focal point for the work of several organizations. Such a federation is in one sense robust—it is no mean feat to generate $100 million in government funding, design and build the tank, fill it with ultra-pure water, and surround it with electronic components that function around the clock to record the effects of subatomic particles passing through the tank. How, we wonder, can it be simultaneously so vulnerable to a downturn in the economy of a wealthy nation? It will not do to suggest that the project personnel are scientific geniuses with no aptitude for politics. There may be scientific genius in the project, but there are boatloads of political savvy and organizational skill as well. Kamioka would not have come this far without an effective organization of people and talented managers of resources.[2]

The standard starting point for coping intellectually with a phenomenon such as Kamioka is the phrase "Big Science," which is more than a label but less than a concept. Since its introduction by Price (1963) and Weinberg (1961, 1967) the term has become a fen of vagueness and ambiguity through overuse. It points to the exponential growth of science in the West, the increasing proportion of social resources devoted to scientific activities, and the large scientific projects established by government laboratories. Now studies of "Big Science" include university-industry collaborations in power production, the allocation of credit in collective efforts, the functioning of space programs, the effects of deadlines and scientific administration, and even the spatial distribution of firms in science parks.[3] "Big Science" also connotes the cooperation of researchers in teams, the massive infusion of funds from the state, the involvement of industry in directing R&D projects, the politicization and militarization of the research process, and the creation of international science and technology networks to address global problems. Our conclusion is that the notion is simply too broad to be analytically useful.

We believe it is essential to set aside our amazement at the scale of such undertakings as Kamioka and focus instead on their organization and on their pursuit of instrumentation. One of the most significant developments in late-twentieth-century science is the frequency with which the appetites of researchers for instrumentation have outstripped the resources of the organizations that employ them. Multi-organizational collaborations have come to populate many areas of science that require complex instrumentation or contributions from distinct disciplines or sectors. Whether the objective is to produce experiments in particle accelerator centers, to build and equip spacecraft or telescopes, or to obtain data from Earth's inhospitable places, the co-presence of scientists, engineers, project managers, technicians, graduate students, and staff from a variety of organizations constitutes a novel development in the social organization of science. In Karin Knorr Cetina's words, they embody new forms of social organization, an "eruption of structures . . . at once global and seemingly communitarian in nature . . . that represent possible alternatives to the small-scale conduct of expertise on the one hand and to other global structures" (1999: 201).

This work is an empirical study of an increasingly important structure for the production of knowledge in the modern world: the multi-organizational collaboration.[4] Whenever new organizational forms proliferate, they raise a variety of questions. How do such collaborations form? Are larger collaborations more bureaucratic than smaller ones? What are their principal means of organization? Are organizational forms related to technology? To what degree are collaborations dependent on trust and other factors? These questions form the skeleton of the chapters to follow.

We have built a database of 53 collaborations by analyzing the transcripts of oral histories conducted for a series of documentation research projects. The American Institute of Physics' Center for History of Physics carried out these interviews during the 1990s. In thematic terms, AIP's concern with documentation was similar in many respects to historical concerns with scientific change and sociological concerns with organizational structure and function. Archivists concerned with how collaboration records are produced, how they are used, and where they end up need to understand the collaboration's evolution, structure, and function. AIP's selection of cases reflects an interest in the policies of archives in the United States. Though some international and European collaborations were included for comparative perspective, our sample largely represents North American science. AIP's efforts have generated material for the first statistical, "meso-scale" examination of multi-organizational collaborations. The scope of the AIP sample is a pragmatic constraint on our research. Insights gained from the qualitative analysis of interviews AIP conducted with participants and administrators enabled us to develop sensible measures of the characteristics of collaborations. Using standard statistical analyses and classic organizational concepts, we identify relationships among the significant structural dimensions of the collaborations. We interpret these relationships by means of specific historical illustrations. These relationships are a map through the maze of possible pathways that scientists pursue to satisfy their curiosities and career ambitions; the histories reveal the mechanisms of modern collaboration. By discerning and then explicating these relationships on the basis of the material at hand, we believe we can both advance sociological understanding of these new organizational forms and enrich policy discussions of their utility and impact. Our interpretation is based on the assumption that *the structure of collaborations is best viewed in terms of the practices through which data are acquired.*[5] Without data, scientists cannot make claims of empirical knowledge and cannot build research careers. What matters most are the forms of organization that integrate or segregate structural elements of collaborations; these forms shape the autonomy of member scientists in producing knowledge.[6] The principle that may be said to emerge is that bureaucracy is the protector of freedom: collaborations use particular organizational forms to define their participant's rights to acquire and use data. We will not argue this point now, but we hope to demonstrate it in the chapters that follow.

In the next section, we distinguish our sense of "collaboration" from a variety of closely related senses. Next, current perspectives on scientific collaboration are reviewed. Because it is somewhat unusual, our specific method of combining historical and sociological techniques is treated in some detail. In the final section we introduce the twin themes of technology and bureaucratic organization, summarizing the line of argument in the five substantive chapters that follow.

Teamwork and Collaboration in Modern Physics

An empirically informed understanding of multi-organizational collaborations as organizations is essential for the sociology of science.[7] For the purposes of this study, an "inter-organizational" or "multi-organizational collaboration" is a research project involving at least three independent organizations.[8] This definition focuses attention on collaborations that are predicated on a federalist organization of power. The projects undertaken by *all* the collaborations in our sample could have been performed by single organizations had the political will existed to provide a single organization with the necessary resources or to empower a single organization to command the resources of other organizations. Although some nations are more inclined than others to create self-sufficient scientific organizations, even the large, territorial institutes of the Academy of Sciences in the former Soviet Union participated in some of the collaborations in our sample.[9] Today, with a single remaining superpower, science is entering a prolonged period in which voluntary cooperation among independent, often competitive organizations will be the primary means of pursuing large-scale research.

Several important modes of collaboration do *not* meet our criterion for multi-organizational collaboration. We are not concerned with cases whereby a single organization develops its research program partly by contracting out for the services of other organizations; the contractors in this scenario are selling their services rather than agreeing to dedicate resources to a common set of objectives. We are not concerned with collections of competing research teams that periodically swap findings or materials. Such groups often view themselves as "collaborating," but they are not coordinating their resources or setting a common agenda.[10] We are not concerned with cases in which one or two organizations are able to assemble the resources to carry out a large research project. Finally, we are not concerned with collaboration in the sense of the informal, unmanaged communication that has linked scientists in networks across organizational lines since the seventeenth century, if not longer.

Our study re-conceptualizes, in contemporary organizational terms, the theme of "teamwork" in scientific research. This staple of sociological analysis became prominent in the study of the sciences when sociologists and physicists recognized that aspects of collective, coordinated research, so effective in the mission-oriented Manhattan Project (Perry 1993: 957), were becoming routine in branches of peacetime research aimed at academic audiences. A. M. Thorndike of Brookhaven National Laboratory, which served as an important postwar organizational model for research both in the United States and in Europe, observed in 1967 that the "experimenter" in physics was rarely, if ever, a single individual:

> [He] may be the leader of a group of younger scientists . . . the organizer of a group of colleagues . . . a group banded together to carry out the work with no clear internal hierarchy . . . a collaboration of individuals or subgroups brought together by a common interest, perhaps even an amalgamation of previous competitors whose similar proposals have been merged by higher authority. . . . The experimenter, then, is not one person, but a composite . . . a social phenomenon, varied in form and impossible to define precisely. One thing, however, he certainly is not. He is not the traditional image of a cloistered scientist working in isolation at his laboratory bench.[11]

Such "teamwork," now held to be an indispensable feature of modern science, became an object of sociological research in the 1960s. The analysis of stratification in science examined phenomena such as the Matthew Effect, an aspect of which was the differential accrual of recognition by renowned senior scientists over their junior colleagues for collaborative work (Merton 1973). Warren Hagstrom, who in his classic 1964 work on the "scientific community" devoted a great deal of attention to group work, argued that the mutual dependence of researchers rendered collaboration necessary.[12] Hagstrom distinguished two forms of collaborative research: the traditional and the modern. The former had as its paradigm the professor-student association. The latter was embodied in a more complex form of organization, involving an intricate division of labor and greater centralization of authority.[13]

Absent from the considerations of sociologists in the 1960s was that continuing growth in the numbers of scientists working on an experiment would require increasing the number of organizations involved in an experiment. The history of particle physics after the 1950s illustrates that increasing the number of researchers on a project increases the number of organizations that are involved. Thus, a typical bubble-chamber collaboration at the European Center for Nuclear Research (CERN) in the mid 1960s consisted of about 15 physicists. One decade later the number of researchers working cooperatively on CERN's largest bubble chamber, Gargamelle, was about 50, from seven organizations. In 1985 the Delphi collaboration, working with the Large Electron-Positron Collider at CERN involved more than 350 high-energy physicists from 37 organizations in 17 countries (AIP 1992d).[14] Particle physics was not alone in this postwar development.[15]

What bear and reward sustained study today are cases in which organizations find their individual interests overlap sufficiently to merit joint pursuit of their ambitions, the competition among the individual participants is internalized and managed within the collaboration, and the contributions of three or more organizations make a project feasible. These larger, more complex, and relatively formal affairs are named and socially recognized nodes of organization within science. Their novelty and prominence revive questions that were posed in the 1960s, when both sociologists and physicists realized that meaningful experimental work could not be done by isolated and self-sufficient scientists: What is the basic unit of scientific research if not the individual? How do individuals build reputations and careers in view of this new organizational framework? How do permanent organizations build reputations and strengths when they must combine resources with competing organizations? What strictures do collaborations impose on their members? What forms of governance do they adopt, and what are the jurisdictions of their governors?

Approaches to Scientific Collaboration

To date, no comprehensive theory of scientific collaboration exists. Yet scholarly studies of collaboration are neither novel nor scarce. However, they are overwhelmingly either statistical studies of authorship patterns in the published scientific literature (bibliometric studies) or detailed narratives of individual projects, organizations, or sites. Though valuable, such studies are limited. To compare collaborations across specialties and to reach empirically based conclusions about their characteristics, evolutions, and outcomes requires the analysis of collaborations from several areas of the sciences. This is the first such study.

Bibliometric studies use the public evidence of research activity—most often papers and reports, patents and agreements—as indicators of trends and processes. Co-authorship is typically employed as an indicator of collaboration. Because scientific journals are specialized by field and because journal articles include the organizational affiliations of the authors, bibliometric studies can determine trends in collaboration across nations and across areas of science. Studies have generally shown that multi-institutional co-authorship has increased over time. For example, the percentage of such articles in the Science Citation Index[16] rose from one-third of all articles worldwide in 1981 to one-half in 1995.[17] This trend is evident for every country and every scientific field measured.[18] While worldwide multi-authorship in physics is actually lower than rates for all fields, the incidence for U.S. physicists is higher.[19] International collaboration is increasing too. From the early 1970s until the early 1980s the proportion of internationally co-authored papers doubled (Luukkonen et al. 1993). Measured as a percentage of all co-authored articles, international co-authorship increased from 17 percent in 1981 to 29 percent in 1995 across all countries and fields. Not only is international collaboration increasing; inter-sectoral collaboration also has grown. About 25 percent of all papers published by academic authors involved co-authors in another sector, compared with 20 percent in 1981.[20] Such results were one motivation for our study. But the available data for bibliometric studies set severe limits on the interpretability of results and on the character of questions that can be asked. Co-authorship stems from several types of social relationships—between colleagues, between teachers and pupils, between supervisors and assistants, and between junior and senior researchers. Yet the nature of these ties is not generally identifiable from the published record, and not all such relationships imply the existence of a formalized collaboration. Nevertheless, the trends are consistent with the view that collaboration among organizations is increasing in frequency and importance. In combination with the high public profiles of collaborations and the anecdotal impressions of eminent scientists regarding the increasing importance of collaborations, the case for sustained effort toward improving our understanding of collaborations is compelling.

The more serious limit on co-authorship data is that such data cannot generate insights into the internal dynamics of collaborations. Bibliometric studies group specific collaborative projects in static snapshots, without any indication of the underlying processes of formation, organization, and outcomes. The published result is the only evidence of a collaboration, divorced from social organization and context. Why did these scientists collaborate? How important was the distinction between leaders and followers? How often did they meet? How did they resolve their difficulties? In short, what happened during the process of collaborative work?

At the other end of the spectrum are theoretically informed case studies of particular collaborations or sites. Historians of science have provided extensively documented narratives of the development of accelerator laboratories that host particle-physics collaborations[21] and have recounted the stories of individual collaborative experiments (Krige 1993; Galison 1997). Anthropologists and sociologists have observed collaborative research projects and have provided important interpretive tools (Traweek 1988; Zabusky 1995; Collins 1998; Knorr Cetina 1999).

Case studies do overcome the main limits on bibliometric research, but they represent a beginning rather than an end. They share a microsociological focus, a qualitative methodology, a cultural-anthropological or narrative orientation, and (owing to the research intensity required by the approach) an emphasis on single organizations, centers, or projects. Their strengths are in providing theoretical guidance, identifying social processes, and raising questions about important organizational and cultural dimensions. But when the findings of case studies are contrasted, they display such diversity as to defy generalization. Three examples may suffice.

(1) Carlo Rubbia's success at lobbying the management of CERN and its political overseers to develop the accelerator needed to search for theoretically predicted heavy particles put him in position to form the collaboration credited for discovering the W boson (Krige 1993). In contrast, Lyman Spitzer, the main scientist to lobby the National Aeronautics and Space Administration (NASA) and Congress to develop an orbiting optical telescope was unable to secure any role in the design and construction of the telescope (Smith 1989: 248–258).

(2) The particle physicists who were building the first time projection chamber reluctantly and resentfully conceded that they had to abandon their role as patriarchal masters of their engineers for a power-sharing arrangement (Galison 1997: 574–640). In contrast, geophysicists planning to make *in situ* measurements in polar regions craved power-sharing arrangements with logistics experts and resented having to take time away from what they considered proper scientific work to master the intricacies of managing an expedition (Schild 1997). In further contrast, the European Space Agency has placed engineers in charge of space-science missions. The scientists contributing instrumentation do not even appear on mission organization charts and must compete with each other to acquire the resources that mission engineers control (Zabusky 1995: 70–102).

(3) Leon Lederman, who led the string of Fermilab experiments that included the discovery of the bottom quark, used his position to encourage his collaborators to tackle the topics he considered significant and to forge a consensus among participants about the quality of individual findings (Nebeker 1994). In contrast, Albert Silverman, the first spokesperson for the CLEO collaboration, was a coalition builder who sought out common and complementary interests among collaboration members who generally preferred to keep their own counsel on the direction and quality of research (Genuth 1992: 99–125).

Qualitative case studies, while illuminating both structural and cultural aspects, are unable to provide a systematic assessment of the relative importance of one process over another. Dimensions such as communication, the division of labor, work as a process, technology, negotiation, and size are all "crucial" to the scholar who discovers their importance, but little attempt is made to show why some may be more important than others. Further, it is not clear whether the collaborations that have been studied are representative. To what extent are the findings of case studies generalizable? Is an observed relationship unique to a particular collaboration, or is it a pattern characteristic of most? Our approach, a "meso-level" comparison of 53 collaborations in the physical sciences, enables us to characterize types of collaborations and to assess the importance of structures on the basis of their connections with processes or outcomes of interest. Far from "everything" being related to "everything else," there are relatively few patterns that emerge.

Method

Debating methods has been one of our favorite pastimes over the course of this study. Reflexivity is inherent in a project that has mobilized historians, archivists, and sociologists at a variety of institutional locations to examine multi-institutional collaborations in physics. That we required a multi-institutional collaboration to study such collaborations seemed to strike at the heart of questions regarding the "how" of the study. The contrast between historical and sociological approaches could always be counted on to arouse passion: the historical imperative of diving into primary materials for a nuanced assessment of particular scientific developments, versus the sociological necessity of understanding events as *kinds* of events. It seemed as if interdisciplinarity could only be achieved by placing two distinct studies side by side, with nothing but proximity to connect them.

But historical methods are located in historians, and sociological methods in sociologists. We began this project working for several years in two distant locations—historians and archivists at one, sociologists at the other. The turning point came when one of the sociologists moved to the other location, cross-fertilizing the teams and accounting for the method eventually adopted. Intellectually, the significance of this move lay in the realization that a relational understanding was an opportunity for all and a hindrance to none. More fundamental than the idea of case study versus statistical analysis, of archival immersion versus reliable coding, of qualitative detail versus quantitative precision, was the notion that our common questions centered on *comparisons* between these new kinds of scientific organization. Were there any commonalities in form and process? How best could we characterize the patterned diversity that was apparent in rudimentary comparisons of these collaborations? We rejected the notion that information about a single collaboration, however elaborately studied, however fully described, would be sufficient to make any claims about the nature of collaboration. As the number of cases mounted, each case became an illustration, but not one that could serve as representative until we could be convinced it manifested a repeated pattern. Arguments about representativeness were our most important source of understanding.

Galison describes the "image" and "logic" traditions as fundamental to the history of twentieth-century physics (1997: 19–31). We found the contrast an apt metaphor for our own methodological difficulties over the kinds of information we should collect, the ways it could and should be examined, and the format in which it should be presented. We came to see it as a window on the struggles in which the science and technology studies (STS) community has been increasingly disinclined to engage. As the field has drawn strength and numbers from outside its 1970s core of scientists and sociologists, the *case study* that describes a particular controversy, technological development, or scientific event, has become the common denominator, the common language of the field. In microphysics, the "image" tradition rests on a "deep-seated commitment to the production of the 'golden event': the single picture of such clarity and distinctness that it commands acceptance" (Galison 997: 22). Such an image would be, in its way, so complete and well defined, so free of distortion and background, that no further data need be brought to bear for the demonstration. In physics, these images were pictures of objects or particles, taken to be the constituents and characteristics of the world, such as Anderson's picture of the positron in 1932 or the bubble-chamber picture of the single-electron neutral-current event in the 1970s. Even when the image presented was the result of a compilation of thousands of other images, the *demonstration* took the characteristic form of the selected picture. In the field of science and technology studies, there are such demonstrations, or golden events: the failure of formal means to transfer the skill of laser building or the indigenous knowledge of Cumbrian sheep farmers in their interactions with radiation scientists (Collins 1974; Wynne 1989).

In the "logic" tradition, a single picture, no matter how free of distortion and background, could never provide evidence of a fundamental kind. That kind of evidence rested on statistical demonstration. In the cosmic-ray experiments of the 1930s, Geiger-Müller counters were placed above and below a gold brick. The particle penetration of the gold was determined by the excess of joint firings of the two counters over the calculated accidental rate. Statistical significance was the important matter, since particular co-incident firings could well be accidental. In STS, too, there have been demonstrations in the logic tradition: the Matthew Effect, showing that rewards tended to accumulate over time to those with early career productivity; the increasing density of social ties with the development of problem areas (Mullins 1972; Cole and Cole 1973).

To reap the advantages of both quantitative and qualitative methodologies, we shifted our approach to the comparative, organizational properties of scientific collaborations. Many of these properties may be viewed as structural, but they are features generated by underlying social processes, and ultimately they result in scientific results, reputations, and success or failure. Because our goal was to understand the structure of scientific collaborations, we sought to create conditions under which simple forms of qualitative and quantitative analysis could be combined systematically. The analysis shifted during the course of the project from the micro to the meso level. The focus changed from interaction and everyday practice to the examination of multi-institutional collaborations as inter-organizational formations. The units of analysis became the collaborations themselves. Instead of studying how *people* interact in scientific projects, we examine how *organizations* work jointly in a collaboration, the structural features of these combinations, their variations, and their patterned social consequences.

What we sought was a combination of these two traditions of Image and Logic, of "Casing" and "Counting." Specific, historical details should illuminate the particular case in the context of a range of like cases. The problem, then, became how to select such cases and how to identify their relevant dimensions. The solution we adopted was to begin and end with the individual case, filtered through a quantitative examination of features that we could examine for all of the collaborations in the study. This involved five steps: (1) selecting a variety of collaborations in the physical sciences, (2) identifying a set of common dimensions, (3) categorizing the collaborations according to their values on these dimensions, (4) cross-classifying dimensions in order to discover significant associations between dimensions, and (5) identifying collaborations that represented these associations. The fifth step was crucial: we sought to determine which associations lent insight into common collaborative processes and which ones were spurious, uninterpretable, or meaningless.

The history of our own collaboration is a fair summary of these steps, involving the evolution of three main data-collection efforts combined in the analysis.[22] These phases move from high descriptive detail but less focused comparisons in a relatively narrow scientific area to more focused comparisons in a broader range of areas with less descriptive detail. One major finding that emerged from the sequencing of phases is that collaborations in high-energy physics—our first field of study—should not be taken as typical of collaborations in general. Subsequent phases impressed upon us the range of variability in the phenomenon of scientific collaboration.

AIP's Center for History of Physics provided the data for our analysis. The AIP Study of Multi-Institutional Collaborations in Physics and Allied Sciences was initiated in 1989 by Joan Warnow-Blewett and Spencer Weart to establish documentation strategies for the identification and preservation of records. The first phase, through the early 1990s, was devoted to the study of particle physics. The second phase examined space science, geophysics, and oceanography—field sciences with long traditions of teamwork. In both phases, we interviewed between five and fifteen members of particular collaborations, selected to cover a range of scientific styles and conditions. Through 1995, approximately 500 interviews had been conducted, focusing on the history and organization of these collaborations, their technologies, their management, and their outcomes. The third phase, which continued through the late 1990s, included 23 additional projects in five new specialties: ground-based astronomy, materials science, heavy-ion and nuclear physics, medical physics, and computer-mediated collaborations.

AIP started with particle physics because it has become both paradigmatic of collaborative work and a leading example of the scientific ascendancy of the United States after World War II. In the 1930s, atomic physicists, most notably E. O. Lawrence at the University of California at Berkeley, developed particle accelerators that required unusually large machinery and buildings, but great European experimentalists such as Enrico Fermi, the Joliot-Curies, and Otto Hahn used the charged particles flung out by radioactive materials either to probe atoms directly or to make neutrons, which could probe atoms without having to overcome the forces of electrical repulsion. However, as a starting point for experimentation, particle accelerators eventually won out over radioactive sources and bigger accelerators over smaller accelerators for physicists determined to investigate atomic substructures. American physicists after World War II used their new-found abilities to raise and manage government funding to create national laboratories that took over the development of accelerators too large, too expensive, or too specialized for universities or corporations to undertake. But since many physicists preferred to work in universities, and since one purpose of particle-physics experiments was to train new physicists, collaborations became standard for particle-physics experiments, with university groups typically contributing instrumentation to detect particles and national-laboratory groups providing instrumentation to customize the accelerator beam for the experiment. European nations appropriated this arrangement when they funded and built the European Center for Nuclear Research.

Around the time the AIP study began, Sharon Traweek estimated that there were approximately 1,000 active researchers in particle physics in the world, with another 2,000 abreast of the most recent developments (1988). In this first phase our selection of collaborations was facilitated by the existence of reliable databases on projects at most accelerator sites, as well as funding agencies such as the Department of Energy and the National Science Foundation. We concentrated on collaborative experiments approved between 1973 and 1984 at five of the major accelerator centers: Brookhaven National Laboratory, the Cornell Electron Storage Ring, CERN, the Fermi National Accelerator Laboratory, and the Stanford Linear Accelerator Center.[23] The experiments varied by detector type, target type, accelerator type, and scientific objective (e.g., a search for a rare process, or a crucial test of a theory). The final selection of collaborations was based on a list of 72 experiments from these sites. The list was first restricted to 27 experiments considered to be most important, and later reduced to 19 on the basis of the availability of participants for interviewing. Since inter-organizational collaborations are defined as those involving three or more organizations, we sought to interview at least one individual from each participating organization, including physicists, graduate students, engineers, postdocs, computer specialists, and technicians.[24]

In particle physics, most collaborations formed around accelerators in one of two ways. In some, a few leaders were committed to the exploration and use of all facets of a particle, a process, or an experimental technique. Such collaborations performed strings of experiments. In others, coalitions of physicists with diverse scientific interests, but with a common interest in a detector or an accelerator, performed free-standing experiments. Although every collaboration had a spokesperson to serve as a link between the collaboration and the accelerator laboratory, the responsibilities and duties of those individuals varied considerably. In string collaborations, the spokespersons were typically physicists who had conceived of measurements that could be accomplished with incremental changes to previous instrumentation designs; in free-standing collaborations, the spokespersons were those who initially suggested novel detector designs that could attract physicists with diverse interests. In the latter, administrative burdens were much greater. These collaborations were more likely to change spokespersons, either to improve their organization or to spread the onerous work among more participants. Hence, collaborations differed in levels of organization, in leadership, in joint planning, and in interdependence, as well as in the technologies of fixed-target and colliding-beam experiments.

The second phase of our study broadened the sample of collaborations to include space science, geophysics, and oceanography. Would these collaborations differ from those in high-energy physics? In space science and geophysics the role of the state is extremely significant; the formation of collaborations is lengthy and political. Industrial contractors were needed to develop instruments that would operate in the field for long periods. A wider range of sectors are involved in these collaborations than in particle physics, but again, collaborations were focused on research sites. In particle physics these sites were accelerators. In second-phase specialties, sites were either research vehicles such as spacecraft and ocean-going vessels or systems of data gathering. In geophysics and oceanography an attempt was made to include seismographical, climatological, and oceanographic research, both internationally and nationally recognized projects, and both smaller and larger collaborations. Fourteen projects in space science and geophysics were selected, most dating from the same period as the experiments in high-energy physics. The final sample consisted of six collaborations in space science and eight in geophysics and oceanography.[25]

As we began the interviews in the second phase, the unique qualities of high-energy physics became apparent: the dimensions of interest had to be generalized or we would not be able to answer the question raised earlier: Is an observed relationship unique to a particular collaboration or is it a pattern characteristic of many? We conducted 219 interviews with academic and government scientists, modifying our instruments owing to the complexity and variety of institutional settings. As in the first phase, transcripts of the interview tapes were obtained and indexed by historical and organizational themes. As an understanding of these fields grew, it became clear that other actors had important roles in space science and geophysics, so interviews were conducted with policy makers and program officers of funding agencies.

In space science, collaborations were managed by government agencies in space-flight centers; in geophysics, collaborations formed with multiple funding agencies and interests. Projects in these areas were field-oriented, with important structural contrasts to particle physics. Often experimental techniques were borrowed from other branches of physics and detection techniques were developed by the military. The

(U.S.) National Aeronautics and Space Administration and the European Space Agency provided management authority for engineers in flight centers, while teams of senior and junior scientists built instruments to meet the engineering constraints of a spacecraft. Because they could deal individually with project managers, the autonomy of individual science teams was high, with a Project Scientist designated to coordinate issues among a group of principal investigators. In some cases a new subcommunity of scientists was created, but only after political campaigns to marshal support within the space agency and the scientific community.

In geophysics, numerous funding agencies exist with diverse goals and structures, such that no standard organizational template is possible. Projects differentiate according to whether they "imported" or "aggregated" techniques. In the "importing" model, capabilities or techniques that had proven useful in other scientific areas or in industrial work were introduced into academic geophysics. Scientists formed consortia and hired executives to manage the development and deployment of instruments with the input of standing committees. In the "aggregating" model, a diversity of experimental specialists was mobilized to investigate a site or a process. A Science Management Office (SMO) was then organized to oversee the collection of data. The director of the SMO, who was usually one of the specialists deploying instrumentation, attempted to balance the specialists' common needs against their desire for individual autonomy. The scale, importance, or complexity of the instrumentation usually constrained the SMO director's options.

The third and final phase of the AIP project shifted away from the collection of exhaustive data on particular cases toward a systematic methodology that would allow the incorporation of a greater variety of fields along a limited number of dimensions. As a project team, we felt we had engaged with a significant number of collaborations and that the dimensions identified would be useful in the examination of collaborations outside these core areas. We sought to describe these new projects using a small number of informants, and after the second phase we were prepared for diversity.

The third phase involved a sample of 23 collaborations in five fields. The first field, which we labeled "uses of accelerators," was chosen to compare how scientists other than particle physicists organize themselves to use particle accelerators ($n = 6$). Ground-based astronomy ($n = 7$) was included because inter-organizational collaborations are prominent in the construction of telescopes such as the Wisconsin-Yale-Kitt Peak Project and the Columbus and Magellan telescopes, and in very-long-baseline interferometry. Materials research illustrates collaborations between universities, industrial, and governmental laboratories, such as the Superconductivity Center, microchip consortia, and the Polymer Interface Center ($n = 4$). In medical physics and clinical medicine ($n = 3$), multi-institutional collaborations are formed to pursue a new technique or therapy (e.g., digital mammography, image-guided needle biopsy). The final field, which we labeled "computer-centered collaborations," is not a traditional field of science, but it seemed clear that this type of collaboration would become more significant in the future because of the increasing ease with which data can be shared electronically. In addition to the Upper Atmospheric Research Collaboratory, which attempted to merge data streams from multiple remote instruments in real time, we included collaborations in parallel computation and cosmology ($n = 3$). What these projects share is a focus on a special mode of work centering on computers.[26]

The development of the survey instrument assumed increased importance as we sought to increase the number of cases by interviewing fewer scientists in each collaboration. The interviews from high-energy physics, space science, and geophysics were thematized in order to determine dimensions of collaboration that cut across all fields we had studied to this point.[27] We eliminated features that, however important in a specific case, seemed idiosyncratic or unique to the particular field of study. The dimensions that formed the basis for the final group of interviews were those that emerged as significant—or at least relevant—for the collaborations previously examined.[28]

Once the data from these interviews were collected and coded, individual records were aggregated to produce the data source used in what follows. Our method required collaborations as the units of analysis. First, information on each of the variables was averaged over the several informants for each project. Next, each aggregated variable was examined in relation to the individual component scores to determine the most reasonable aggregate score. Aggregate file variables were recoded to reflect the closest approximation to this "best summary" of the opinions of those involved in a particular collaboration.[29]

After we were satisfied with the coding and analysis of the third phase, we incorporated a group of collaborations from the first two phases by using the same set of dimensions. Clearly, our ability to do this was predicated on the fact that the third-phase instrument was developed from the dimensions initially identified from the interviews from the first two phases.[30] We revisited the original interviews and coded 110 interviews on

30 collaborations from the first two phases.[31] Owing to missing data, an attempt was made to follow up with 30 of those interviewed earlier.[32] Hence, the empirical analysis is based on information from participants in 53 scientific collaborations.

We began by exploring these thematic dimensions in the interview transcripts. Next, we examined univariate and bivariate relationships between these factors and other measured dimensions in this group of collaborations. Although such a procedure is guided by preliminary and revisable notions of likely associations, in the end it is largely inductive. It is not worth wasting too much time on ideas that do not pan out. The challenge arises when a relationship is evident between two dimensions that might plausibly be related. Many associations do not represent large differences, either statistically or substantively. Others can be interpreted, but not without an imaginative stretch. Since the historical understanding of cases *precedes* the search for statistical relationships, many associations were rejected as uninteresting.

The pivot point of our approach is the relation between the specific and the general, the illumination of a particular case in the context of a range of "like" cases and the illumination of a relationship in the context of a specific set of historical details from particular cases. Bar charts are used to display evidence for certain kinds of relationships. These, however, have another important function as a *selection criterion* for our interview extracts. How do most scholars and ethnographers select illustrative quotations? One hopes they are selected for some representative purpose. Yet with a large number of interviews it is no exaggeration to say that *interview extracts could be marshaled to illustrate almost any relationship*. To take one example from our analysis, it would be inappropriate to select materials that speak to the importance of trust from collaborations that saw themselves as successful. Why? Because such material can equally often be found in others that do not view themselves as successful. That, indeed, is at the root of what it means to say that trust is not related to assessed performance, as chapter 5 shows.

After identifying those associations that seemed promising or revealing, we returned to the case histories of individual collaborations that *constituted* the association. Take, for example, a positive relationship between conflict and interdependence. Such a relationship is positive because many conflictual collaborations are highly interdependent. Identifying these collaborations and examining the reasons for the association in particular cases lends specificity to the analysis and renders it less likely that the correlation is spurious. For example, in the discussion of the relationship between conflict and interdependence in chapter 5 we identify an experiment at the Stanford Linear Accelerator Center that helped us to understand the projects in which conflict between scientists and engineers proved more likely to occur.[33]

Implementation of the method across the five main substantive chapters involves the identification and analysis of specific cases that represent particular types of collaboration (chapters 1 and 3), or particular relationships (chapters 1, 2, 4, and 5) in order to understand the processes that characterize these social organizations in a more concrete fashion. Each chapter contains at least five illustrations of varying lengths. The main fields (particle physics, space science, geophysics, ground-based astronomy) are used at least twice. Seventeen cases, about one-third of the total employed in the quantitative analysis, illustrate such features as formation, magnitude, organization, technology, and conflict. About half of these illustrations begin in one chapter but recur in others, though in deference to the reader these are adjacent in all but two cases. The collaborations we frequently use as exemplary are Fermilab 715 (particle physics), Voyager (space science), DND-CAT (materials science), AMPTE (space science), and Keck (ground-based astronomy).[34] Other cases, including IUE, CRPC, and STCS, are used to illustrate a simple relationship. Every chapter includes at least one brief illustration from particle physics—one of our main themes is that this field is relatively unusual.[35] The case studies provide some insight into common collaborative processes, and they help to determine which patterns to interpret as meaningful and which to reject as uninterpretable or as beyond our imaginative powers.[36]

The weakness of our method is the selection bias inherent in a sample of collaborations that actually came to fruition. The results presented here must be viewed in the light of the kind of collaborations included in the study, a sample of 53 relatively successful collaborations, at least in the sense that they persisted long enough to receive resources. If a collaboration began and collapsed, or was discussed but never funded, it was not included in our range of cases. Simply put, we are unable to say anything about collaborations that were not collaborative for long. Quite possibly, such beginnings of organization constitute a larger number of scientific occasions than those that persist. Much remains to be said about collaborations from the time an idea is put forth until the day of funding or rejection.

Overview

Why collaborate? Put so starkly, the question implies a direct answer: Collaboration is viewed as the best or perhaps the only way of achieving one's objectives. Why have collaborations become more common and prominent in scientific communities? Because individual organizations cannot command the money, facilities, and expertise needed to acquire the kinds of data their scientists find meaningful. Thus, collaborations are appropriately described as "technoscientific," a term that gained currency in the 1990s as a way of emphasizing the fuzzy boundaries between equipment, practices, inscriptions, and claims in the local contexts where knowledge is created (Bijker et al. 1989). Collaborations are technoscientific in the specific sense that concept, design, and organization all revolve around the technological practices required to collect data.[37] Our central argument is that an understanding of modern scientific collaborations requires close consideration of the shaping roles of technology and bureaucracy. In theoretical terms, knowledge of nature is the "moral object" of collaboration, but the design and acquisition of instrumentation is the "real program," the proximate goal (Wuthnow 1987).

Just as large scientific collaborations must develop technological practices for acquiring data, they are also predicated on developing an organizational structure. There would be no point in discussing collaborations without some characterization of their organizational properties, whether from a scientific, a humanistic, or a managerial point of view.[38] Unfortunately, characterizing organizations for readers outside the social sciences is greatly complicated by the connotations of the term "bureaucracy," which many natural scientists consider an epithet or pejorative. Its use condemns a certain type of organization without argument. We will nevertheless employ the term "bureaucracy" in its traditional, descriptive sense to mean a hierarchical organization with a well-defined division of labor and with formal rules or procedures for achieving goals (Weber 1946). There is no better term that comports with both scholarly and ordinary language.[39] To say that one collaboration is more bureaucratic than another is not to criticize the first and praise the second; criticism and praise begin by asking whether a collaboration has the right level of bureaucracy for achieving its goals under its circumstances.[40]

This book is organized in five substantive chapters concerning the formation, the magnitude, the organization, the technological practices, and the experiences of scientific collaboration. How are collaborations formed? What differences are associated with larger size or longer duration? How are collaborations organized? Is there a relationship between organization and technology? What features, if any, are associated with the success of collaborations in the eyes of their participants? In the pages that follow, these questions are addressed through a combination of statistical analyses and case histories that illustrate a variety of structural relationships.

We begin in chapter 1 with the process of forming large projects. Using cluster analysis, we identify five basic ways collaborations form. These forms do *not* correlate closely with research specialty but with the level of complexity required by the collaboration. We introduce the term "encumbered" for collaborations that faced resource uncertainty, pressure from parent organizations, structural change at the funding agency, or the involvement of an external authority in selecting participants. The degree of encumbrance affects the organization and the interdependence of collaborations, with lasting consequences for their development. Collaborations tend to be larger when they form in the context of pressure from funding agencies and parent organizations. Owing to the importance of size in earlier discussions of "Big Science," chapter 2 focuses specifically on the magnitude of inter-organizational collaborations. We use "magnitude" in preference to "size" because the temporal dimension of collaborations, particularly duration from idea to funding, is as important as the number of participants.[41] Whatever the particular arrangement of members, the magnitude of collaboration has implications for its organization and management. Larger collaborations, not surprisingly, tend to involve greater formalization and more hierarchical structures. But by limiting the scope of activities subject to formal procedures and hierarchical decision making, even bureaucratic collaborations manage not to violate their members' sense of the quintessential nature of science. For example, where matters directly concern the production of results, deliberations are often widely participatory and decision making is often overtly democratic.

Chapter 3 focuses on the variety of ways in which collaborations organize themselves to accommodate the sensibilities of their individual and organizational members, to comply with the requirements of their funding agencies, and to satisfy the managerial and administrative perquisites for acquiring meaningful scientific data. The principal finding is that multi-organizational collaborations display patterned diversity, but the patterns do not generally coincide with scientific specialty. There is no "geophysics style of collaborating" or "materials science style of collaborating." Particle physics is an exception to this generalization, but even these collaborations can be used to illustrate contrasting categories and relationships. *Bureaucratic* collaborations possess many of the classical Weberian features: a hierarchy of authority, reliance on written rules and regulations, formalized responsibilities, and a specialized division of labor. *Leaderless* collaborations, like bureaucratic ones, are formally organized, highly differentiated structures. Yet in contrast to the more bureaucratic collaborations, these projects did not designate a single scientific leader to represent the interests of scientists or to decide scientific issues. The third principal form is *non-specialized* collaborations, which possess lower levels of formalization and differentiation than bureaucratic and leaderless collaborations. But because non-specialized collaborations, like bureaucratic collaborations, have a hierarchical structure for scientific decision making, we consider both non-specialized and leaderless collaborations "semi-bureaucratic." The remaining collaborations are *participatory*. Their members describe decision making as wide open and consensual, define organizational structure through verbally shared understandings or legally non-binding memoranda, and have few levels of authority and little use for formal contracts. Because particle-physics collaborations dominate this category, we speak of "particle physics exceptionalism."

Scientific collaborations are fundamentally dependent on equipment. Technology, broadly defined as the set of instruments and practices that scientists employ in the acquisition and manipulation of information, embodies interdependence and autonomy within collaborations. Chapter 4 considers how technological practices are related to organization. In general terms, the level of bureaucracy in collaboration is inversely related to the level of interdependence in the acquisition and analysis of data. The purpose of creating bureaucracies in these collaborations was to ensure that the organizations or teams would be autonomous in acquiring or analyzing data, whereas the purpose of participatory organization was to ensure that data streams would be collective property.[42] Most collaborations pursue partially interdependent technological practices, adopting aspects of bureaucracy that ensure their members can operate autonomously in some areas but must reach consensus in others.

These issues are crucially important for the scientists that participate in collaborations because they lead directly to credit and conflict that affects careers. In research where one's time, instrumentation, and ultimately data are one's own, there are no such issues. But no work on the sociology of recent collaborations should be considered complete without an effort to understand how collaborations are experienced by participants. Chapter 5 delves into perceptions of success or failure, with particular emphasis on their relation to trust. Trust has been a persistent theme in the literature on both science and organizational process. Social studies of science have ascribed a central role to trust in the constitution of knowledge since the beginning of modern science (Shapin 1994), yet there are few studies of its operation in collaborations where actors must coordinate their efforts toward a common goal. We find that foundational trust—trust that other scientists can contribute to the joint enterprise—*is* important for multi-organizational collaborations. But this foundational trust is so widespread that its presence does not discriminate between more or less successful projects. Assessing other forms of trust, we found no relationship between trust and the success of collaborations. Nor is there a relationship between complex trust and the extent to which projects are built from pre-existing relationships.

Why is trust thought to be so important? We find that trust is inversely related to conflict—which is, in turn, positively associated with bureaucracy. This explains why trust is generally viewed as positive. As much as scientists might like to conduct research in collaborations with colleagues they know and trust, the reality is that they are often in collaborations with strangers. They have become adept at developing mechanisms for overcoming deficits of trust and managing conflict. What we call the "paradox of trust" is that bureaucratic organization segments work to impose a structure for interaction that is in some sense non-collaborative. By elaborating formal structures for social practices, collaborations minimize mutual dependencies and reduce the need for high levels of trust. Some interdependence is characteristic of all inter-organizational collaborations, but the close interdependencies, low bureaucracy, and fluid organization that characterize particle physics are atypical.

Most of the large-scale projects analyzed here have technological motivations, are difficult to initiate, require massive efforts from their principals, and involve decision making with imposing career consequences. In all the other fields we examined, scientists in collaborations were more independent than particle physicists in the generation and dissemination of scientific results. They were also more autonomous in the activities that constitute the groundwork of collaboration. Technology, broadly conceived, is the basis for collaboration. The independence so valued by scientists is resistant to fracture *because* of bureaucracy, not in spite of it.

1
Formation

The institutional framework of modern science includes a host of organizational actors: funding agencies, academic departments, national laboratories, corporate laboratories, government laboratories, nongovernmental institutes, and private donors. Their interactions sometimes produce collaborations. Origins are intrinsically interesting. But stories of origins—what Knorr Cetina (1995) calls "birth dramas"[1]—are also significant because they inform participant understanding, both during the project and after it has dissolved. The most convincing evidence of the power of these narratives comes from our interviews. No matter what kinds of prompts one uses, no matter whether the purpose of the interview is to collect quantitative or qualitative data, no matter whether the interview is long or short, the most extensive set of remarks by our informants was on the subject of project formation. Narrative logic dictates that one starts an account of collaboration by a story that explains how it came to be.[2]

Psychological necessity is one thing; predictive power is something else. We characterize the origins of collaboration and search for patterns involving how they originate and how they end up conducting business. Three steps are involved. First, in the opening section of the chapter, we review the main contextual issues facing scientific collaborations and conceptualize the dimensions involved in forming projects. These include the degree of uncertainty in the acquisition of resources, the way projects fit into established programs, their association with organizational changes, whether previous research provided a point of departure, and their institutional composition. In the second section, we use cluster analysis to identify five ways collaborations form and illustrate each with a description of one representative collaboration. These descriptions are an opportunity to observe concretely what is meant by an inter-organizational collaboration as well as to indicate the diversity of these arrangements. In the third section, we examine statistically significant relationships between formative processes and other features of collaborations. We focus on resource uncertainty, institutional origins, and the role of participating organizations as they condition the evolution of the collaboration.

Our goals in this chapter are to describe the logics that underlie the formation of collaborations and to analyze how the formation of a collaboration shapes its evolution. We ask where the impetus to form a collaboration originates, where participants are located structurally, and how groups and organizations are incorporated. Those instigating collaborations generally start by asking what kinds of instrumentation and resources their projects need and who has the power to decide whether they go forward. Social identities ground the mechanism of project formation firmly in its social environment, including the organizational actors that produce knowledge in the modern world. The fundamental reason for the lasting effect of the project-formation stage is that the identities of the leading constituents are typically established at the outset. For the majority of collaborations, major changes in the identities of leading institutions or individuals are rare.

The formation of collaboration involves the establishment of linkages among social actors, a process that also entails *de-linking*. The organizations, people, and groups that collaborate must be drawn away from other structures and interests, those with alternative claims on time, activities, and resources. At the micro level, of course, we refer to individual scientists and the close-knit and long-lasting work groups associated with these scientists in a team or a laboratory group. Researchers always enter a collaboration at some specific point in their career trajectories. Even in graduate school this involves a choice (attractive or coerced) to devote one's time to *this* project rather than another. At any subsequent point in one's career, the movement into a collaboration is not only a choice to do something; it also involves opportunity costs—choices *not* to engage in alternative activities. In the project-formation stage, individual scientists and teams must be dislodged, dis-embedded from ongoing projects, and committed to new activities. Such commitments are thought to be easier if they involve relationships with other scientists who are known and trusted, in projects that involve the renewing of old collaborative partnerships or the opportunity to work as colleagues with those who have sometimes been friendly competitors.

Dimensions of Project Formation

The impetus to form collaborations may be construed as the interplay of factors arising from four broader dimensions of the environment: (1) the inter-personal context (the relations among independent scientists whose employment terms allow them to choose their research problems), (2) the funding context, (the availability of state or private patrons and the fiscal and political climate in which these patrons operate), (3) the sectoral context (relations among academic, industrial, and governmental sectors, which traditionally have different goals and cultures), and (4) the context of participating organizations (relations among established, permanent organizations—university departments, national research laboratories, government research laboratories, corporate research laboratories). Table 1.1 summarizes the distribution of project-formation characteristics of multi-organizational collaborations.

Inter-Personal Context

Elementary considerations lead to the hypothesis that pre-existing relationships among some or all of the participants are central to forming a

Table 1.1
Distribution of project-formation dimensions ($n = 53$). Unless otherwise noted, all coding is dichotomous.

Variable	Mean or percentage
Interpersonal context	
Pre-existing relationship[a]	2.42
Brokered relationship[b]	1.68
Donor context	
High resource uncertainty	19%
Partial or full reorganization of funding agency	34%
Sectoral context	
University only instigated	13%
Dominant sector	57%
University dominant sector	50%
Geographical dispersion of instigators[b]	2.11
Home organization context	
Pressure from home organization	35%

a. Scaled 1 = low, 3 = high.
b. Coded 1 = regional, 2 = national, 3 = international.

collaboration. This simple sociological hypothesis was confirmed by many of our informants. For example:

> We've known each other a long time. We haven't ever been scientific collaborators, but we're old friends. I'm not old friends with everybody that was involved in the partnership but I sort of know them all. Friends of mine are friends of theirs or something. . . . The relationships are pretty close.

Note that this "standard approach" to collaborations as resulting from ongoing social relationships among scientists tends to obscure other formative processes. It should not be assumed that collaborators always seek collaboration:

> None of us . . . have any experience with collaborations, and by nature of the personalities of the people involved I think that they are not—I include myself in this—are not people that would be in a collaboration unless we were forced into it. We're forced into it by a couple of circumstances. First of all there is declining funding in science in general. And also . . . the very strong preference by the program director at the [National Science Foundation].

Therefore, in addition to asking informants about their pre-existing relationships with participants, we also asked about "brokering"—that is, cases where members of a collaboration are brought together by other parties.[3] Emergence via brokering and pre-existing relationships need not be mutually exclusive. Brokers can, in principle, bring together friends or even competitors in a scientific field—what particle physicists refer to as "shotgun marriages."[4] Officials of funding agencies, one informant made clear, are especially well situated to broker collaborations:

> There were other proposals in the same area that were being formed and put together at that time, and one of the competing proposals came from X . . . when those two proposals were brought to the NSF the people at the foundation recognized that instead of funding one of these centers or funding neither, they had an opportunity perhaps to talk to the leaders of these proposals and they made an attempt to merge the two proposals.

We assessed "inter-personal context" by probing the breadth and depth of collaborators' pre-existing relationships and the extent to which they sought each other out to form the collaboration.[5] We rated the importance of pre-existing relationships "high" when major participants in a collaboration had previously worked together in research or administrative duties.[6] We rated the importance of brokering "high" when a funding agency or other external authority actively determined who became members of a collaboration.[7]

In our sample, collaborations are more likely to form around pre-existing relationships than around brokering by external authorities (table 1.1). However, brokering is found in every discipline we studied except ground-based astronomy, and is most prevalent in geophysics. It should be viewed not as an aberration but rather as a standard technique for forming collaborations. In a few cases, both brokerage and pre-existing relations were rated "high," as when an authority insisted a single collaboration be formed from two collaborations that had each formed through pre-existing relationships. But in general, emergence through brokering and emergence through ongoing relationships are inversely related. Collaborations brokered through funding agencies are less likely to involve a group of participants with strong pre-existing ties.

Funding Context
Two features of the funding context constitute significant sources of variability for inter-organizational collaborations: resource uncertainty and agency reorganization. These are two sides of the relationship between the nascent project and the resources that might make it possible. Each feature refers to the level of entrepreneurship required to launch the collaboration—from the perspective of the would-be collaborators in the former, from the would-be donors in the latter.

"Resource uncertainty" indicates whether or not a collaboration has a "manifest" source for the resources it needs. A "manifest" source indicates a funding agency with a program that readily fits collaboration objectives or a research facility with the capability to support the investigations. "Manifest" does not mean "easy." Competition for manifest resources can be brutal. For example, American particle physicists know that they can readily apply to established programs in the Department of Energy or the National Science Foundation for funding for experiments or capital improvements, but the requests for funding may far exceed the program's budget. In contrast, a group of universities with ambitions to construct an astronomical observatory encountered high resource uncertainty whey they sought private benefactors and entered a plot line associated more with melodrama than with science:

> There was a woman in Los Angeles who was quite wealthy, who loved her husband and wanted to create a memorial for him [by funding an observatory] and who herself was dying of cancer. . . . There was urgency here because her health was failing and the very evening on which they [the woman and university officials] were to sign this [the papers bequeathing funds to the collaboration], she was rushed to the hospital and she died . . . without actually nailing the money in

place. There was the evil witch sister on the scene who descended immediately and the net result was that we never saw that money.

In this case, the would-be collaboration faced no competition from other scientists, but instead had to deal with the contingencies of an individual's estate planning.

"Reorganization," usually of a funding agency, indicates whether collaborations with manifest funding sources were predicated on the creation of new programs or on arranging for support from multiple sources. The latter is represented by international collaborations that often form to combine support from multiple governments to pursue projects that overtaxed national sources. An example of the former is a medical-physics collaboration that came together because of a funding agency's adaptation to a changing budget environment:

> At the time, the budgets of the Defense Department were being cut, the Cold War had ended, and so I guess some savvy person in the Army said "Women want more funding for breast cancer research, our budget is being cut, [and] if you don't cut our budget as much, we can allocate money for breast cancer research." And the Army has always had a medical research program, so this was added.

There is probably no better way to stimulate researchers to reconsider their working relations than by creating a new funding category in which to compete.

The level of resource uncertainty distinguishes collaborations that justified themselves by claiming to be superior within a recognized research niche from collaborations that went hunting for donors with arguments that their plans made novel or under-appreciated research possible. Collaborations with greater resource uncertainty emphasized the novelty over the competitive excellence of their plans. "Funding-agency reorganization" distinguishes between collaborations that required entrepreneurship to form and collaborations that formed in the course of affairs that seemed standard to the scientists involved. Entrepreneurship, whether from the funding agency or the petitioning research scientists, was required for collaborations predicated on funding-agency reorganization.

Sectoral Context
The distinctiveness of modern collaborations lies partly in the inclusion not only of multiple organizations but also of multiple organizational forms, institutions, or sectors. A variety of institutions are represented in this group of collaborations: (1) university departments, (2) university

research institutes not affiliated with any single academic department of a university (e.g., oceanographic institutes), (3) government-funded research institutes that are free standing or independent of any university base, (4) government research laboratories, and (5) corporate research laboratories. By separating the first and the second of these, we distinguish collaborations in fields with strong links to traditional educational curricula from those in fields with weak links. We use an undifferentiated category of "research institutes" for all manner of government and government-funded research organizations because their various legal arrangements were generally not relevant to their roles in collaborations.[8]

Our primary distinction is whether or not the collaboration was instigated by a particular type of organization. A collaboration had a dominant instigating sector when its original membership was overwhelmingly a single type of organization or when participants from one sector pointed to institutions in another sector as having taken the lead in designing the collaborative project. The alternative occurred when project planning was diffused among institutions from multiple sectors. Space-science projects were particularly complex in this regard. Some were conceived and incubated within the space-flight centers of NASA or ESA, some were generated outside the flight centers, and some were conceived and incubated in a combination of advisory panels and flight-center studies.[9] Overall, universities were plainly the most important sector in our sample, but they were dependent on other sectors (table 1.1). This dependence is consistent with historical findings on the development and importance of scientific research in industry, government, and other organizations.

Organizational Context
Each collaboration in our sample included participants employed by three or more organizations. However, there was wide variation in the extent to which organizational interests figured in the formation of the collaborations. In some instances, prospective collaborators needed explicit, non-trivial approval from their organization to participate. Occasionally, prospective members were unable to join the collaborations because of their failure to secure that approval. When corporate scientists needed to convince their supervisors that collaboration membership was consistent with protecting their firms' intellectual property, or when university physicists needed to convince their departments, a dean, or other officials to support the collaboration, the role of the parent

institution was relatively large. In contrast, collaborations of university physicists whose activities were covered by ongoing contracts between their organizations and a funding agency were rarely affected by pressure from their home institutions.

A contrasting form of pressure is to *join* collaborations.[10] This most obviously occurs when a project is extremely large and an organization, for a modest investment of resources, can gain access to a facility it would otherwise be unable to use. In other cases, organizations view collaborating as a means to sidestep obstacles to funding sources, as when for-profit corporations use relationships with universities to justify petitioning for public funds:

> It's not just that you collaborate or have these research teams to stick your heads together . . . but these days collaborations are often formed to obtain funding, and this is really changing. . . . Many of the science programs at big American companies . . . are shrinking. In the past universities have sometimes looked to big companies to get fellowships or grant for their students or post-docs. I think it's completely changing, that scientists and companies are trying to get money from the U.S. government by working through collaborations with universities and national laboratories, because it's somewhat difficult for a company like X to obtain a direct government grant. . . . They want their own scientists to move up and work more on technology related things. . . . The science problems are farmed out to universities. Everybody moves up the value chain. Value in this case is technology. It's money, making money.

About one-third of the collaborations in our sample dealt with some form of pressure from parent institutions. Pressure from parent organizations was more likely to occur in space science and geophysics, especially as compared with particle physics and ground-based astronomy. The major facilities of particle physics and astronomy—accelerator laboratories and observatories—have often been intended for the ongoing use of outsiders. Neither they nor the employing organizations of the scientists that use them need to make accommodations for their collaborative use. In contrast, the facilities of geophysics and oceanographic institutes are often intended for the use of staff rather than outsiders. An institute or an outsider's organization frequently needs to bend its customs or rules for a collaboration to take place.

Five Patterns of Formation

Attending to funding, sectoral, inter-personal, and organizational contexts enables us to examine the nature of the obstacles to forming a collaboration. Geographic dispersion among the instigators, when coupled with obvious funding sources and sectoral homogeneity, may indicate that collaboration among institutions was normal for a particular branch of research. In fact, this combination describes most particle-physics collaborations. Sectoral heterogeneity coupled with strong pre-existing relationships and a reorganized funding source might indicate that collaborating across sectors was unconventional. This combination describes many collaborations in materials science. Systematizing this approach, cluster analysis[11] is employed below as the basis for the identification of five general patterns of collaboration formation. In the third section of the chapter, an analysis of relationships between these contextual factors and other dimensions provides a foundation for a discussion of the influence of a collaboration's formation on its evolution.

A cluster analysis sorts objects according to their similarities across a range of dimensions. Here the objects are scientific collaborations, while the dimensions are variables related to the four contextual factors above. Cluster analysis on selected project-formation variables[12] yields a graph, or dendrogram, that groups collaborations on the basis of how closely they resemble each other across these variables (figure 1.1). By drawing a vertical line through the graph, we divide the collaborations into clusters. Note that the further to the left one draws the line, the larger the number of clusters but the smaller the number of collaborations in each cluster. For these clusters, the members are quite similar—up to the limit where a single member constitutes a cluster. The farther to the right one draws the vertical line, the smaller the number of clusters and the larger the number of members. In these larger clusters, the members are less similar to each other—but still more similar than to members of other clusters.

There is a tradeoff between the intellectual economy of a small number of clusters (at the far right of the figure) and the substantive cohesion of highly differentiated clusters (at the far left). We selected a five-cluster solution for its balance of generality and interpretability.[13] The combinations of variables that distinguish these five clusters are our basis for characterizing the contexts that give rise to collaborations.

(1) Dominant Sector/Conventional Collaboration ("Wake Up and Smell the Coffee")
Suppose you are an established scientist with good connections to comparable scientists who work for the same kind of organization. And further suppose you realize that there is an excellent opportunity to launch a research project that would *not* stretch the norms of what your home

34 Chapter 1

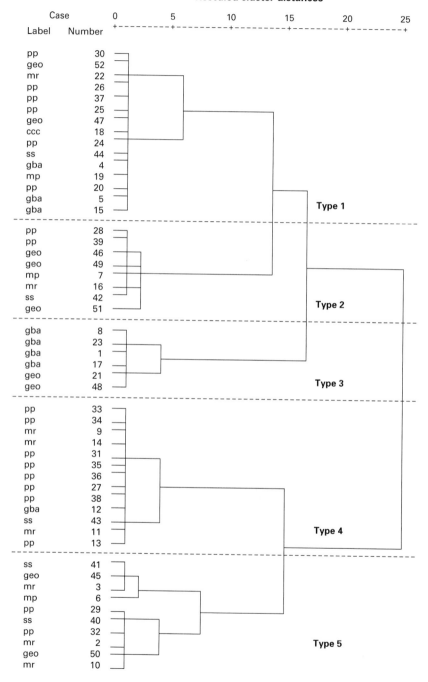

organization expects you to do. But to take advantage of the opportunity, a collaboration must be formed. What do you do? The obvious answer is that you find an appropriate time to tell your colleagues at these other organizations to get busy—this is an opportunity ripe for the plucking! Once you have enough such colleagues to appear credible as a collaboration, you can take steps to attract the additional participants you need.[14]

That, in the abstract, is the story of the formation of the collaborations in this group. They formed because members of similar organizations agreed that they had an excellent shot at performing prominent research if they proposed a project that required they work together. The impetus for starting these collaborations varied—a boldly imaginative idea, changes in the available facilities for research, the need to learn new skills in light of technological change, expansion in the horizons of ongoing research, or the creation of new funding programs in government agencies. But in all cases, someone spread the word among peers from similar organizations while parlaying his dissemination of the opportunity into a dominant position for his organization in the collaboration's formation. In terms of the specific variables that created this cluster, these collaborations were distinguished by an absence of pressure from the parent organizations of their members, by a dominant sector, by an obvious source of funds, and by their inclusion of a mix of scientists who had and had not worked together previously.

Take as a concrete example the Active Magnetospheric Particle Tracer Experiment (AMPTE). Since 1958, when James Van Allen and his colleagues at the University of Iowa discovered the radiation belts that bear his name, the source and dynamics of the belts have been an obvious

Figure 1.1
Dendrogram showing project-formation clusters. (Type 1: Wake Up. Type 2: Full Speed. Type 3: Get Out. Type 4: Resources Wanted. Type 5: Come Together. pp: particle physics. ss: space science. geo: geophysics. mr: materials research. gba: ground-based astronomy. mp: medical physics. ccc: computer-centered collaborations.) To derive the clusters we used Ward's method, a commonly used procedure for cluster linkage which is based on minimizing the sum of squared within-cluster distances. These distances are calculated as squared Euclidean distances and for each case the squared distances to the cluster means on all variables are computed and then summed up. Thus, cluster membership for cases is attributed according to the smallest increase in the within-cluster sum of squares. "Rescaled cluster distances" refers to the horizontal scale which indicates at what distance clusters are combined (the greater the distance the less the similarity). Distances are rescaled to fall in the range between 1 and 25 for standardization and easier reading.

source of scientific questions with relevance to the problems of operating satellites. In 1971, Tom Krimiges, a Van Allen protégé who had taken a position at the Applied Physics Laboratory (APL),[15] realized that if the solar wind "blew" charged particles into the magnetosphere, then the process could be investigated by releasing tracer ions outside the magnetosphere and searching for those ions inside the magnetosphere. His initial calculations convinced him that detection and release capabilities were adequate to demonstrate the existence of such a process at energy levels it could plausibly be assumed to occur.

After his epiphany, Krimiges ran his ideas past Erwin Schmerling, the NASA Headquarters scientist with responsibility for magnetospheric physics.[16] Krimiges also promoted his idea during the opportunities for relaxed shop talk surrounding an invitation to speak at the Max Planck Institute for Extra-Terrestrial Physics (MPIET) in Germany. Like APL, MPIET designed and built spacecraft to investigate near-Earth space; unlike APL, ion releases were a specialty of MPIET. The wisdom of pursuing this idea collaboratively was as obvious to one American as the features of a face: "I remember Hermann Föppl, who was their chemical expert on how to do this [release ions], and when I met him, . . . I noticed that his eyebrows had been burned off in one of those experiments." Krimiges's counterpart, Gerhard Haerendel, was receptive and supportive. That news was sufficient for Schmerling to begin a trickle of funds that enabled the organizations to study designs for mechanically coupled spacecraft for ion identification and ion release. Krimiges and Haerendel also lined up fellow scientists, inside and outside their organizations, whose instrumentation they wanted to include in a full-blown proposal to build the spacecraft—making sure in the process that the spacecraft also provided these scientists with the ability to use their instruments for less risky explorations of the natural environment. Most notably, the German scientists successfully argued for the inclusion of British colleagues, who turned the inert mechanical adaptor between the German and U.S. spacecraft into a third spacecraft.[17]

The collaboration never wondered where and how to obtain funding; only when there would be an opportunity. While MPIET could commit to providing its spacecraft,[18] all the prospective participants understood that APL needed dedicated NASA funding orders of magnitude beyond what Schmerling controlled. Matters languished until NASA requested that outsiders propose projects. But when it did, Krimiges and Haerendel were ready to draft a proposal that balanced the risky active experiment against the more secure goals of observing features of the natural environment and addressed all the technical pitfalls that their engineers had been able to find. The external panel NASA assembled to review the proposals recommended AMPTE be funded, and the collaboration was in business.

Because AMPTE was not conceivable without the capabilities of the Applied Physics Laboratory, the Max Planck Institute for Extra-Terrestrial Physics, and the Rutherford-Appleton Laboratory (which built the British spacecraft), research institutes were clearly the dominant sector at the outset. Because AMPTE's means and ends fit readily into these organization's traditions, the parent organizations did not seek to influence the project. Because AMPTE was space-based research into theories of magnetospheric dynamics—not Earth-based research and not research into environmental effects on military satellites—it was obviously appropriate for NASA to consider and not appropriate for either the National Science Foundation or the Department of Defense. And because the collaboration needed to expand to diversify its goals and broaden its appeal, its major participants included both scientists who had not previously worked with any of the others as well as scientists who had previously built and instrumented spacecraft together.

(2) Dominant Sector/Unconventional Collaboration ("Damn the Torpedoes—Full Speed Ahead")
Now imagine you are a scientist sensing you are at a crossroads in your career. Maybe you have recently been hired and need a project that will make an impact on your new employer. Maybe you believe you are near the end of what you can do with your current resources and skills and have your eye on research techniques that you do not have the means to access. Maybe the political economy of research is squeezing your discipline in a way that behooves you to consider pursuing research differently. And further suppose you receive the message that a collaborative venture has the potential to secure you a new niche. Do you dismiss the opportunity because you don't know and may not be compatible with the other people or organizations involved? Do you dismiss it because participation will likely require that you spend time fighting to change some rules or modify some traditions in your organization, or fighting to change the rules or modify the traditions of someone else's organization, or will likely require that you compromise your personal or organizational interests for the greater good of the collaboration? The people in these collaborations did not. They ignored the torpedoes and managed to form collaborations before the explosions sank their prospects.

These collaborations formed because scientists and their employing organizations had to work together to revive the scientific prospects of their scientists. In the majority of cases, there was no individual whose ideas were clearly responsible for the project. In others, an individual's ability to control resources was more important than intellectual creativity for bringing together the collaboration. In most cases, conflicts among the participating organizations or their main participants were the source of torpedoes. In others, change in a single organization of strategic importance to the collaboration was necessary for the collaboration's existence. As was the case with those in the preceding category, these collaborations had a dominant instigating sector and an obvious source of funds. Unlike that category, their members had generally not worked together previously, and their members' parent organizations influenced the collaboration because the collaboration promised (or threatened, depending on one's point of view) to change the relationships among the parent organizations.

The Incorporated Research Institutes for Seismology (IRIS) is a particularly dramatic example. The political economy of seismology was disheartening to many university seismologists in the early 1980s. Those interested in monitoring the Earth's vibrations at frequencies relevant to discerning large-scale, internal structures of the Earth were chafing at their dependence on government seismic observatories for detecting underground tests of nuclear explosives. Those interested in the regional structure of parts of the Earth envied oil-exploration geophysicists, who acquired huge data sets by recording the vibrations of large numbers of intricately planned explosions and then filtered the data for information that would determine whether oil deposits existed at accessible depths. And every seismologist could walk into the office of a scientist in any number of other fields and see how new computers and more sophisticated programming were generating unprecedented capabilities for acquiring and analyzing digital data.[19] Though a fortunate few had the social and technical abilities to work with national security or industrial data, or had the resources to develop instrumentation, most American university seismologists felt the need for an initiative that would enable them to build and analyze their own large data sets.

At various conferences, university seismologists lamented the state of their research resources and built momentum for a pan-university initiative. Independent workshops held in late 1983 and early 1984 to lay out the substance of a National Science Foundation proposal for the development of a digital global seismic network and for digital portable seismometers were well attended and intellectually successful.[20] When word filtered down from NSF that two large seismology proposals could undercut each other's chances, the two groups agreed to form a single consortium and drafted a joint proposal, which NSF agreed to fund.

The workshops also provided participants with evidence of the conflicting interests they would be internalizing in forming a collaboration. Some seismologists wanted a university that had developed its own seismic station instrumentation to be the collaboration's headquarters. Others feared that the collaboration would not treat members equitably if managed from one of the participating organizations. Seismologists from universities with good relations with government agencies wanted to build on what was already in place. Others sought autonomy from government programs. Seismologists from universities with strong computing centers wanted the collaboration to concentrate its resources on instrumentation. Others wanted the collaboration to provide support for the processing of large data sets. Seismologists from universities with laboratories for instrumentation wanted the collaboration to support their efforts. Others wanted the collaboration to tap industrial suppliers of instrumentation to the government and oil industry. How IRIS was structured and functioned was bound to affect the powers and relative standings of the universities with Earth Sciences departments that hired seismologists.

Significantly, several scientists who were important in figuring out an administrative structure for the collaboration felt they were entrusted with administration because they had *not* been among the instigators and were thus not partisans in the conflicts that surfaced at the workshops. They consulted university counsels, who made sure the consortium's articles of incorporation and by-laws ensured its tax-exempt status but did not address issues of collaboration governance. The scientists adopted a modestly revised version of the by-laws of an extant Earth Sciences consortium and hoped for the best. The worst nearly came to pass. At one of the first meetings of the board of directors, disgruntled members planned to move for an immediate change in the by-laws in order to force a reconsideration of the collaboration's balancing of interests. The chairman of the board, having heard of the plan, brought along a corporate lawyer (retained at the collaboration's expense) who "explained the facts of life of this corporation" to the members of the board. The move was defeated, and no one has since threatened outright revolt over an intra-collaboration controversy.

IRIS had to represent academic seismology as a whole to justify its ambition to obtain large quantities of digital instrumentation and to develop data-management techniques for the enormous data sets such instrumentation would create. The homogeneity of IRIS's sectoral composition did not make its formation easy. Its instigators could not restrict collaboration membership and remain credible; consequently, most participants had not previously worked together. The collaboration could not discuss the characteristics and availability of the instrumentation and techniques it hoped to develop without addressing its impact on the distribution of resources for seismology among American universities. Thus, IRIS's member organizations, unlike AMPTE's, were suspicious of each other, and each tried to use collaboration procedures to its advantage.

(3) Entrepreneurial Funding Collaboration ("Get Out of Your Business-as-Usual Rut")
Now imagine that you are an accomplished scientist who, while not exactly at a crossroads, is nonetheless uneasy. Perhaps you have a position at an organization with aging facilities and fear that you will have difficulties remaining competitive with colleagues at other organizations. But when you think about seeking government funding for recapitalizing your organization's facilities, you conclude that there is no way a government agency will spend so much money on anything that is not a national facility. And when you think about broaching the subject to your organization's executives, you conclude you will just give them a case of "sticker shock." Or perhaps you have a secure position at an amply equipped but insecure organization that is under pressure to justify its continued existence. Or perhaps you are at a secure, amply equipped organization but have a structurally insecure position (e.g., a non-teaching faculty position at a university). In all these instances, business as usual will not serve your interests. It will not lead to the acquisition of new facilities, a better justification for your organization's continued existence, or a better justification for your employment within your organization.

Forming a collaboration, for the projects in this category, was a way to get out of the business-as-usual rut. If neither the government nor your parent organization will recapitalize your facilities, form a partnership with a few institutions to share expenses and benefits. If you need to diversify the areas your organization covers with its first-rate facilities, collaborate with those struggling to tackle a promising area with second-rate facilities. If you need to justify your value even though you do not perform or directly support one of your organization's basic revenue-generating tasks, form a collaboration that links your research (and the research of your disciplinary colleagues at other organizations) with the interests of other organizational colleagues who are engaged in revenue-generating tasks.

Doing business unusually carries the burden of developing a new set of procedures, which are tasks best carried out with people with whom you work well. Unlike the collaborations in the formative process discussed above, all these collaborations were formed by people who had worked together previously, and none had an obvious source of funds. Some managed to fund themselves; some combined self-funding with government funding; some raised money from private philanthropies; some prodded multiple program offices of an agency to negotiate a joint arrangement for providing funding; some aggregated funding from several governments. These collaborations did resemble the previous formation in that their promoters faced pressures from their home institutions. Some dealt with opposition from colleagues with different ideas for recapitalizing facilities; some felt obliged to renegotiate their job descriptions in order to feel comfortable taking on the work the collaboration would entail; some had to argue the merits of particular individuals or organizations for intra-collaboration administrative authority. Overcoming such difficulties inevitably created a residue of bruised feelings.

The Berkeley-Illinois-Maryland Array (BIMA), a radio-astronomy facility the three universities built in Hat Creek, California, embodies many of these forces and conditions. A drive to recapitalize an aging university radio observatory could founder even before it left the Astronomy Department because of competition between optical and radio astronomers. An appeal to a national funding agency had to demonstrate why developing a particular university's facilities was in the national interest, especially since the National Radio Astronomy Observatory, founded in 1956 over the objections of astronomers who feared their field would be better served by dispersed funding to individual researchers, had eventually succeeded in carving a niche for itself in the development of facilities that university astronomers could compete to use (Needell 1987).[21] So when Berkeley's radio astronomers heard that University of Illinois astronomers would be interested in jointly upgrading and expanding Berkeley's array of radio antennae, they jumped at the chance: "Even if we gave up half the use to another group, we'd still be miles ahead. . . . It was very much a win-win situation." However, a Berkeley-Illinois proposal to NSF still asked for more than NSF was willing to provide. Thus, when Leo Blitz of the University of Maryland (who had been a postdoc under Jack Welch, the director of Berkeley's Radio Astronomy Laboratory) put out feelers about the prospect that Maryland might join, Welch responded "Sure, it just takes money." With Maryland's radio facilities on the verge of being deemed obsolete, its optical astronomers having no concrete plans for recapitalization, and its new Dean of Science and Engineering controlling some discretionary capital, Blitz was able to parlay his relationship to Welch into an institutional commitment from the University of Maryland.

The addition of Maryland necessitated that the three organizations revise the division of labor and rights of use that Berkeley and Illinois had contemplated. Welch's suggestion that Maryland just needed money to join was effectively a prediction that such matters would not be bones of contention. Indeed, they were not, though some awkwardness was involved in the division of labor. As the builder of original Hat Creek antennae, Berkeley was well equipped to provide additional antennae and hardware while Illinois, as the host of a supercomputer center, was well equipped to provide software and data storage. Maryland took responsibility for the cables connecting the antennae into an interferometer—the hardware Berkeley was least versed in—and the interferometer operations and calibration programming—the software with the most intimate interfaces with the hardware. Additionally, because of the contrasting rules and regulations of three state universities, it took the university counsels a year to work out an agreement that fit the organizations' practices and embodied the understanding reached by the radio astronomers. Nevertheless, with the three universities contributing the needed capital, NSF was willing to provide the additional funds needed to support operations and maintenance for a Hat Creek array that could perform interferometry at millimeter wavelengths.

Collaboration was the price Berkeley, Illinois, and Maryland's astronomers paid to continue to administer a competitive facility in radio astronomy. The collaboration was university-dominated because the point was for universities to run an observatory. The three organizations had the common structural elements of aging facilities and modest capital. However, the principal scientists at each organization relied on know-who rather than an impersonal search for similarly inclined organizations in order to form BIMA. Because they entered into the collaboration with confidence in each other's reasonableness, they took in stride the novel difficulties of dividing labor and coping with differences in their organizations' policies. The gravest threat to the collaboration's formation came from the competition for capital within each of the organizations. During the year of negotiations that produced BIMA's charter, one of the members lost a prestigious optical astronomer who was displeased with the emphasis on radio astronomy. Changing business as usual did have its price, but the radio astronomers did not bear much of it.

(4) Business-as-Usual Collaboration ("Collaborators with Resources Wanted")
Now suppose you are a scientist in a field where it is difficult for anyone to do much of anything *except* in a collaboration, and you have a bright idea for a project. You would contact your professional friends, but probably find that too many were already committed to other projects to participate fully in yours. Then what? Or perhaps you are a scientist who has spent your career in research with a small knot of colleagues, and you have a bright idea for a project that can only be carried out through a sizable collaboration. What do you do? For both cases, the answer is *advertise*: COLLABORATORS WANTED; experience preferred, access to needed resources (e.g., time, money, students, engineers) required. While such ads have never, to our knowledge, been placed in a scientific journal, they are effectively spread by word of mouth.

The collaborations in this group formed because a scientist(s) with a (shared) vision succeeded in selling the vision to enough scientists with the needed competencies. They differ from the first formative logic in that no sector dominates the collaboration, and the collaborators are less likely to have worked together previously. These factors imply that the instigators routinely went beyond their circle of professional friends in order to acquire enough participants to be credible. They differ from the second and third logics in that they do not have to address conflicts between or within participants' parent organizations. The relative absence of parent pressures implies that these collaborations are more common for the careers of the individual participants and their parent organizations. Particle physics, which is almost always pursued in collaborations, is disproportionately represented in this category. Individual particle physicists, their employers, and their funding agencies have all learned to welcome collaborative research.

Take, as an example, Fermilab 715, one of a string of fixed-target experiments that used a beam of hyperon particles.[22] At Fermilab's inception, Joseph Lach moved from Yale to join the Fermilab research staff and began discussing hyperon experiments with his former colleagues. Soon they were personnel-limited. What they could do—potentially even their right to deploy instrumentation at Fermilab—depended on what they could sell others on doing. The obvious indicator that the collaboration core was selling a high-quality product was that it could attract physicists performing similar experiments at other laboratories as opposed to attracting physicists who needed a collaboration to join. Providing an idea appealing enough to convince such physicists to come to Fermilab was the junior physicists' route for upward mobility within the collaboration and, potentially, within their parent organizations.

Peter Cooper, a junior professor with hopes of winning tenure at Yale, came up with such an idea. He realized that the Yale-Fermilab detector could be economically reconfigured to perform a more elegant, definitive version of an experiment that Roland Winston of the University of Chicago had led at Argonne National Laboratory. Winston's experiment had produced results that contradicted accepted theory (Keller et al. 1982). Cooper's idea hooked Winston, who did not consider himself wedded to his Argonne results. The crowning coup was that Cooper's idea hooked the University of Leningrad physicist Alexei Vorobyov, who had been working at CERN on experiments that directly competed with Fermilab.

Winston and Vorobyov had no experience working with Lach, Cooper, and their circle of colleagues. Incorporating them into the collaboration was nevertheless straightforward. They both had ongoing financial support that was sufficient to cover their contributions to the collaboration.[23] Each pursued a strategy for the primary task, electron detection, that was not already covered by the core of the collaboration. Once Cooper and Lach were able to convince Fermilab's director that they were covering all the pitfalls the director's advisory committee could spot, they had permission to build the experiment. The hoped-for results were doctoral dissertations for Yale and Chicago students, tenure for Cooper, and international acclaim for Lach, Winston, and Vorobyov should the anomalous results hold and their choice to pursue hyperon physics be justified as a route to fundamentally new physics.[24]

This collaboration, unlike the entrepreneurial, relied on a core of multi-sectoral expertise: accelerator laboratory physicists to provide a customized beam of particles and university physicists to provide detector components that, in combination, could characterize the processes the beam particles caused or underwent. Forming a collaboration was a necessity for the research careers of Fermilab 715's participants. Each assumed that he (or she) could cope with any competent physicist whose scientific skills and interests were compatible with his own. Except for the uncertainties surrounding visas and travel rights attendant to Soviet-American relations in the 1980s, nobody worried about each other's standing or circumstances within their parent organizations. A memorandum of understanding sufficed for the participants and the Fermilab administration to be confident in everyone's seriousness of purpose and dedication.

(5) Externally Brokered Collaborations ("Come Together, Right Now, Over Me")
Finally, imagine you are a scientist enjoying a satisfying career with a secure position in a secure institution, and thanks to forces largely beyond your control, you and your disciplinary colleagues are all contemplating an obviously stellar research opportunity that begs for collaboration. Perhaps a new facility is coming on line and needs people to make use of it, or a funding agency has started a new budget line and is soliciting proposals, or a rarely occurring natural phenomenon is soon to occur. You may have some misgivings about pursuing such opportunities, because so many other scientists will be interested in participating that it may seem well nigh impossible to field a winning proposal by organizing or joining a collaboration of professional friends in similar organizations.

But if you and prospective collaborators do not come together in a timely fashion over this opportunity, you personally—your scientific community generally—will be the poorer for not having your ideas and abilities in the mix of possibilities. Two possible routes to participation confront you: (1) Prepare as self-sufficient an individual proposal as possible, hope a central authority decides to include you in its mixing and matching of individual proposals, and accept working with whomever is selected. (2) Seek out different kinds of scientists from those you would normally work with, convince them that a collaboration will best serve everyone's individual interests and create an appealing environment in which to form future individual interests, and hope that the socially unconventional character of your collaboration will make you appealingly different from the competition.

As a group, collaborations in this category are less likely to be built on pre-existing relations between collaborators because of the use of impersonal competition to select participants in some cases. These collaborations are less certain of finding funding niches in which to compete for funds because of the unconventional combination of organizations that have different traditions or restrictions for acquiring research funds. These collaborations never have a dominant sector, reflecting the breadth of prospective participants for these widely recognized research opportunities. Finally, these collaborations are more likely to be pressured by their member organizations because of the difficulty of justifying unconventional combinations of organizations.

One collaboration that came together when a central authority mixed and matched self-sufficient proposals was the Voyager project. Scientists at the Jet Propulsion Laboratory (JPL), who made a point of looking for opportunities for planetary-science projects, had alerted NASA Headquarters that there would be a rare planetary alignment that would make it possible, in the winter of 1976–77, to launch spacecraft that could reach multiple outer planets by using gravitational boosts to accelerate the spacecraft from one planet toward the next. JPL engineers and panels of the National Academy of Sciences debated how to upgrade planetary spacecraft for a longer mission that would expose a spacecraft to many environmental hazards while leaving sufficient money, weight, and power for a suite of scientific instruments. When NAS, JPL, NASA, and NASA's political authorities reached an understanding of the spacecraft's engineering parameters and the project's overall budget, virtually every planetary scientist in the world had ideas for what measurements should be made. NASA organized numerous "working groups" to investigate various instrumentation possibilities and to set specifications within which NASA could conduct competitions to design and build the desired instruments. In some instances, working groups became fraught with Byzantine politics as members sized up each other as potential competitors and fretted over how the working group's recommendations would affect their chances in the coming competition. However, by including some competitors in the working groups, NASA ensured the consensus recommendations would be sufficiently generic to conduct an open competition.

The competition was held, and the results created enough awkward social relations to frustrate any conspiracy theorist intent on finding a mastermind behind the selections. "I would say you could have knocked me over with a feather when it was selected," a co-investigator on an underdog proposal recalled. He attributed the selection less to the merits of the proposed instrument's design than to the track record of the principal investigator's organization in turning out well built space instrumentation and the willingness of other experienced scientists to serve as co-investigators on the experiment. A JPL science team's proposal was rejected in favor of a similar instrument proposed by outsiders, who then had to work with JPL's engineers and managers to integrate their instrument with the rest of the spacecraft. Though no interviewee gave any indication of expecting or receiving anything but appropriate professional behavior from other participants, the collaboration was sealed not by a memorandum of understanding but a set of legally enforceable contracts among NASA Headquarters, JPL, and the principal investigators and their organizations.

Another example in which scientists reached beyond their circle of professional friends to improve their chances of participating in a stellar research opportunity is the DuPont-Northwestern-Dow Collaborative Access Team (DND-CAT) that operates at Argonne National Laboratory. The U.S. Department of Energy's decision to build at Argonne an Advanced Photon Source, a new accelerator to generate synchrotron radiation,[25] stimulated all synchrotron radiation users to rethink their goals and working relationships. Partly because of his home organization's geographic proximity to Argonne, the ambitions of Jerome Cohen of Northwestern University increased dramatically. He had been working at a Brookhaven synchrotron facility with disciplinary peers from other organizations on a modestly instrumented beamline that supported their specialized investigations. But he wanted for Argonne an elaborately instrumented beamline that would cost ten times as much as the one he had used at Brookhaven and would support a bevy of various investigations. His Brookhaven collaborators did not consider such grandiose ambitions within their means, so Cohen suggested that DuPont (with which he had a consulting relationship) should want to finance and use an elaborately instrumented Argonne beamline.

DuPont, of course, had its own capital to invest. When it decided that the breadth of its research interests justified applying with Northwestern to Argonne for space for a beamline at the new accelerator, Cohen had to find capital to remain credible. That proved to be a scramble. He did not have the kind of pan-departmental enthusiasm to make Northwestern's administrators consider tapping the university's endowment, though he had been constantly selling synchrotron experiments within Northwestern since construction on the beamline began. Accelerator construction had cost the Department of Energy's materials science programs so much that they did not seem eager for instrumentation proposals; and NSF's materials science programs did not seem eager for proposals to work at a DOE facility. NSF eventually came around, but matters were dicey for Cohen:

Fortunately at that time the State of Illinois came to our rescue, and the Argonne people, [who had] convinced the state that as part of their technology initiative to put up some money for a variety of things [including] some money to be peer reviewed to be given to those universities in Illinois who were participating. And they set down a number, and we got a number which was 450,000 for our particular group. . . . So now we had a seed. We had enough money to be honest with the DuPont people. And the university still hadn't spent a nickel on this.

Still, the funding pressures were such that Cohen and DuPont's research department, which had to absorb funding cuts, welcomed the interest of the Dow Chemical Company, which offered to pay 20 percent of the beamline costs for 20 percent of its availability for research. A collaboration of two competing corporations and one university is rare by the standards of our sample, but then it's not every day (or even every decade) that research organizations get to use the construction of a new accelerator as an opportunity to reconsider their research strategies and their institutional relationships.

The main issue in the formation of both Voyager and DND-CAT was not *whether* there would be a collaboration but *who would be in it*. The research opportunities presented by the alignment of the outer planets and the construction of the Advanced Photon Source were not to be missed. Whether a funding agency mixed and matched proposals to form the collaboration or an instigator sold the idea of collaborating on an unconventional combination of organizations and individuals, the result was a collaboration in which the participants had not previously worked together and in which no sector was dominant.

The Legacy of Project Formation

Important consequences follow when funding agencies or other central authorities applied pressure to form collaborations, when collaborators searched for patrons, when the source of funds was a new funding program, when organizations outside academia instigated collaboration, and when parent organizations sought to influence participation in a project. In each case, the complexities and regulations imposed on collaborations at the outset—that is, their "encumbrances"—affected their organizational structure.[26] When the formation of a collaboration is encumbered, participants should expect certain developments to follow. To illustrate, in what follows we examine the most significant relationships between brokerage, resource uncertainty, instigating sector, pressure from parent organizations, and other aspects of collaborations.[27]

Brokered Collaborations

An external authority is instrumental in "brokering" a collaboration when it informs scientists of their common interests and urges they work together—generally by signaling a positive attitude toward a jointly proposed project—or insists they work together by contractual requirement.[28] Three factors were frequently associated with the application of such pressure: (1) Collaborations became larger than average both in

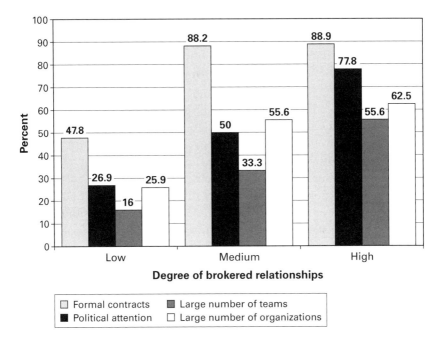

Figure 1.2
Brokerage by size (teams and organizations), formal contracts, and political attention.

terms of participating organizations and project teams. (2) Formal documents were drafted to specify the obligations and privileges of the collaboration's participants. (3) Collaborations attracted political attention (figure 1.2).[29] These features are especially prevalent in space science, in geophysics, and in medical physics. Why should this be the case?

The Voyager Project was used to illustrate how a central authority can insist that a particular set of researchers collaborate. The experts in orbital mechanics at JPL realized that a spacecraft launched in late 1976 could use a rare planetary configuration to travel to multiple outer planets. JPL managers alerted NASA Headquarters of the opportunity and JPL engineers began to contemplate spacecraft designs. NASA Headquarters could have passively allowed JPL's scientists and the outside scientists with good connections to JPL to propose integrated payloads of scientific instruments. Instead Headquarters assembled working groups to prescribe the kinds of instruments that should fly on the spacecraft.

The recommendations of the working groups became the basis for impersonal competitions that NASA ran for the right to design, build, and operate each type of instrument. NASA's implicit decision was that all members of the planetary sciences community should have an equal chance to participate in a fly-by of multiple outer planets. Ensuring an open, meritocratic selection of participating teams and ensuring that as comprehensive a range of scientific interests would be included in the project became more desirable than allowing potential participants to determine who their best collaborators would be. Because Voyager aimed for comprehensiveness, it ended up with more teams than most projects in our sample, and each team was itself multi-organizational to include experts on the instrument and each planet the spacecraft would encounter.[30] Because the participants did not build a mutual understanding of each other's needs and ambitions in the process of drafting proposals, but instead wrote proposals on the basis of the spacecraft's capabilities to support their particular instrument, formal contracts between each team and JPL were desirable to discipline each team and JPL to hold to its commitments.

Central authorities were also frequently involved in unconventional collaborations that had a single dominant sector, as exemplified above by the Incorporated Research Institutes for Seismology. The collaborations in this category were often based on widespread discontent among scientists with the resources available to their research specialty. In IRIS's case, this discontent had a governmental focus in the frustrations of university seismologists with what the Department of Defense and the Department of the Interior were providing. NSF program managers made clear in exploratory conversations with university instigators that they required a consensus, pan-university proposal and that multiple, contrasting proposals would compete with each other to the likely detriment of each. Those instigators (and their universities) who hoped that they could nevertheless build a cozy collaboration around the sophistication that a relatively small number of seismologists had acquired were outvoted at inclusive meetings of all the interested universities. This collaboration could not form except by being large enough to attract political attention and formal enough to survive the conflicts it internalized.

Two independent factors account for the tendency of central authorities to be active in forming collaborations in space science, geophysics, and medical physics. The first is the scarcity of golden opportunities for projects in field sciences. On those rare occasions when predictable natural events intersect with human curiosities and capabilities, everyone will want a piece of the action; a large collaboration will be needed to ensure comprehensive coverage, central authorities will need to ensure that participants were selected fairly, and formal contracts will be needed to specify the terms on which participants who did not select each other will work together. The second is the proximity of some sciences to high-profile public policies. When scientists in these fields seek to change the way they do business, their patrons will insist they produce a community-wide consensus proposal to impress on public officials the unanimity of their discontent. Formal contracts will be needed to give the scientists and their organizations processes for dealing with their intra-collaboration conflicts. When public officials want scientists in these fields to change the way they do business, they will form large collaborations to spread the practices they wish researchers and their organizations to adopt and use formal contracts to hold the (potentially reluctant) organizations and individual researchers to their responsibilities. Space science and geophysics are strongly influenced by the scarcity of golden opportunities for research; space science, geophysics, and medical physics are all strongly influenced by their closeness to high-profile public policies.

Resource Uncertainty
Resource uncertainty—that is, whether or not there was a well-known, well-established program that regularly provided the resources for collaborations in a particular area—is particularly important among the dimensions of project formation.[31] When collaborations have a manifest source of funds and do not face the burden of having to search for sources of support, they are unlikely to have formal contracts, centralized decision making over collaborative functions, or to employ hierarchy in making substantive decisions. But they also are more likely to subject topics for data analysis to collaboration control and to share data widely within the collaboration (figure 1.3), subjects to which we will return in chapter 4.[32] This combination of collaboration self-governance and a broad domain of activities is the distinctive characteristic of particle physics. Obvious funding sources for collaborations are indicative of a community in which multi-organizational collaborations are so common that participants are comfortable with allowing collaborations to regulate many activities without formal specifications and written rules.

Fermilab 715 (a particle-physics collaboration with manifest sources of funds) and BIMA (a ground-based astronomy collaboration that required entrepreneurship and luck to fund) illustrate opposite ends of this spectrum. Fermilab 715 researchers, as has been typical of participants in particle-physics experiments, needed to work with data from multiple detector components to reconstruct events from the variety of

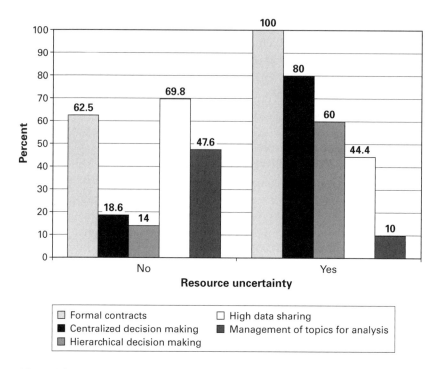

Figure 1.3
Resource uncertainty by formal contracts, centralized and hierarchical decision making, data sharing, and topics management.

particles into which unstable, heavy particles decay. This condition and its consequences were so common and well accepted that the ongoing Fermilab collaboration working with hyperon beams added the University of Leningrad group, with whom none of the Americans had previously worked, because of Leningrad's unique mastery of "transition radiation detectors" (TRDs), which were then the most efficient and capable means of detecting electrons and ascertaining their properties. Even when Fermilab's director told an American physicist on the experiment, to "get rid of the TRDs," the physicist felt entitled to respond "'I think it's an important part of the experiment'" and to ignore "one of the few times that he [the director] told me to do something directly."

By itself, electron detection was meaningless—it had to be combined with information that other components acquired about the identities and characteristics of other particles created in the experiment. The collaboration could not operate meaningfully except by sharing data; and each member wanted the experts on the various components to examine how the data from each was analyzed and used. Participatory and consensual decision making about data acquisition were thus the norm. Any decision the experiment's titular leader made in the absence of consensus was open to reconsideration:

We had some great fights over just how should we set the trigger [to filter out unwanted signals before they were recorded]. . . . [One physicist] and the spokesperson had a great fight in public, where the spokesperson finally had to say "No, we don't change the trigger the way you suggested." The other physicist blew up. The next week we changed it his way. So, what else is new? He was right as usual. But he had to make the case to the satisfaction of the collaboration. This was a week of running. We only took data for six weeks during this experiment. It was a fairly contentious issue.

Communal data acquisition and shared data streams also imposed on the collaboration the need to manage data analyses to ensure due recognition for the member organizations. In one instance, involving dissertation students at two different organizations, the collaboration carefully carved differentiated topics for their dissertations without making its most important results hostage to student inexperience:

Basically no graduate student really published the first paper . . . that was the primary focus . . . where the major thrust of the analysis was. In general you really like students to publish things that they really did the analysis on. The way we broke it up was [one student] took the magnetic moments . . . [which] was a completely orthogonal topic [to the primary focus]. So, that was easy. [Another] student took the asymmetry parameters, having measured all the particles. . . . That had never been done before. That was . . . a transition to [the third student] who had to do the full blown analysis . . . and all that complicated stuff. There was reasonable differentiation. We only had three theses to make. Had we had to invent the fourth, we could have had one guy publish the electron asymmetry and one guy publish the neutron and neutrino asymmetry. We didn't have to do that, but we could have.

A formal agreement would have at best been useless and at worst an impediment for this collaboration. There was no need to codify a division of labor for the construction of detector components because the members were content to proceed on the basis of their previously acquired expertise and the availability of previously used instrumentation. Any advance specification on the acquisition of data or division of topics among students would have interfered with arguments on data-acquisition strategies and the use of students' proclivities and talents as factors in decisions over dissertation topics. In particle physics, manifest funding sources are associated with a well-known culture of collaborating that obviates the need for formalities and makes a norm of regulating data acquisition and analyses on a collaboration-wide basis.

In contrast to collaborations in particle physics, BIMA originated in a context of uncertain resources. BIMA's formation required that radio astronomers at three state universities obtain capital funds from their universities *and* draft a successful proposal to NSF. With three universities spending their own money for BIMA, it was a foregone conclusion that there would be a formal contract stipulating fiscal obligations. The collaboration was required to ensure that each university's interests were served. Overseeing the project were two committees: "the BIMA board, which consists of three people from Berkeley, two people from Illinois, and two people from Maryland. . . . That group decides the major thrust. The day-to-day decisions are made by the executive committee, which is the directors of the three institutions, and that's why we have phone calls every two weeks, and at certain times more frequently." Thus, decision making was centralized and hierarchical on matters that were deemed appropriate for the collaboration as a whole. At the same time, BIMA was organized in order to limit the scope of collaboration decisions and leave meaningful power within the member organizations. Most important, instead of a particle-physics-style collective debate over the best scientific strategy, in BIMA the participating organizations retain significant control over the scientific use of the collaboration's instrumentation. Individual scientists were made responsible initially for generating research plans:

> Everybody who wants to do a project writes a scientific proposal and then we have referees, one from each university and two people from outside, to evaluate what goes on. There is a formal guarantee that a minimum of 30 percent [of use for science] goes to Berkeley, a minimum of 20 percent to Maryland, and 20 percent to Illinois, and then there is publicly available time [reflecting the financial contribution of NSF], but we can compete for that as well.

Data taken with BIMA instrumentation are not shared within BIMA (unless successful proposers strike their own agreements to do so), but they are archived for future use by any interested researcher.

By making the uncertainty of resources a challenge to be overcome rather than a limit on their ambitions, the Berkeley, Illinois, and Maryland radio astronomers got out of their business-as-usual ruts. Through comparison with collaborations like Fermilab 715, with its obvious source of funding, it is clear that BIMA's participants assumed a distinctive set of burdens and constraints. Extraordinary business could only be conducted if certain prerogatives of the home organizations were respected. There was to be little or no redistribution of resources across the established

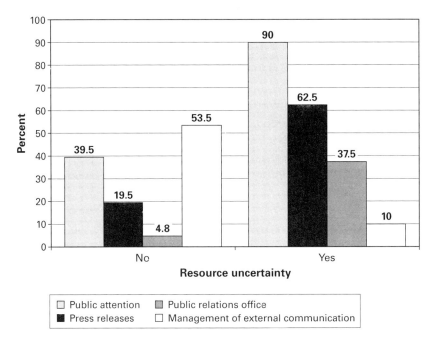

Figure 1.4
Resource uncertainty by public attention, press releases, public relations office, and management of external communication.

organizations. No matter how much better one university's proposals for BIMA's use was than another's, the quantity of time each university had was written into their formal contract and protected by its representatives on the proposal review panel. There was not to be muddling of lines of affiliation of BIMA users. Each university was to have clear grounds on which to judge and reward its users. The scientific leaders of the member universities had to assume responsibility for a centralized, hierarchical organization that did not control very much. Indeed BIMA was more centralized and hierarchical than most collaborations we studied in order to ensure that it would not control any more than was necessary to build the array of radio antennae, to make them operational, and to generate good publicity for the member universities.

Resource uncertainty was also accompanied by heightened publicity and lowered management of external communication by the collaboration as a whole, as illustrated in figure 1.4 and as considered further in chapter 3. Collaborations encumbered with resource uncertainty in their formation were eight times more likely to have a public-relations office, three times more likely to have press releases, and more than twice as likely to attract public attention than projects with obvious funding sources. Collaborations with uncertain resources were also far less likely to impose collective management over scientific publications and other public presentations of results. These features represent an ongoing effort to maintain a parent organization's support for these collaborations by generating benefits (good publicity) without infringing on a parent organization's prerogatives.

After overcoming the difficulties of getting three universities plus NSF to finance BIMA jointly, BIMA made sure that its activities led to good publicity for its member organizations. Participating scientists cooperated with several newspapers, public television programs, and the graphics experts at Illinois' National Center for Supercomputing Applications, which "was interested in getting some publicity about what it was doing, and this is one of the projects that it's supporting. So they financed it. They got a regular professional team to do it, a good producer." Press releases about BIMA from the member organizations were frequent.[33] But BIMA's administration was careful to leave the production and reporting of scientific results in the hands of the member universities and their individual researchers. Users of BIMA instrumentation report their results without involving BIMA participants who were not directly involved in collecting the data. Collaboration-wide authorship on scientific papers is rare.

In contrast, when there is a clear source of funding, as in particle physics, there is less need to publicize the collaboration outside the scientific community and greater collaboration-wide management of publication of results. The only time Fermilab 715 garnered publicity was when a Soviet participant defected to the United States—an event that threatened rather than furthered the collaboration's existence. Scientific claims were subject to collaboration-wide discussion before submission to a journal and author lists of articles included everyone who had been responsible for operating the detector during data runs. The only time a parent organization's interests intruded on the production of results was when the collaboration confronted the desire of students for first-author recognition for papers based on their dissertation research.

Reorganization of Funding Agency
Collaborations sometimes acquire resources from a newly formed office or program within a funding agency. In almost all instances, the agency's reorganization was not a response to receiving proposals that did not fit

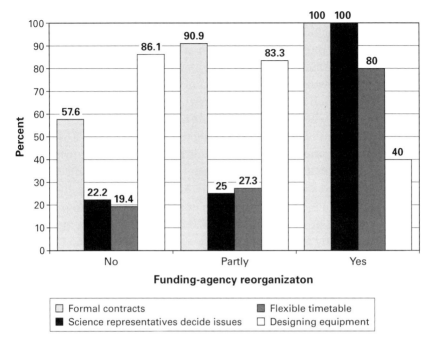

Figure 1.5
Funding-agency reorganization by formal contracts, science representatives, flexible timetable, and designing equipment.

comfortably into its extant categories. Rather, it was an effort by the agency to stimulate novel proposals. Funded collaborations typically proposed that an unusual combination of researchers or organizations coordinate their efforts. Such collaborations invariably operated under formal contracts and made scientific representatives[34] of their organizations responsible for deciding scientific issues; they also tended to operate under flexible timetables and to shun designing instrumentation (figure 1.5).

Recent enthusiasm for involving university scientists in the investigation of materials with technological promise has spawned collaborations that exemplify this combination of characteristics. After World War II, American universities set up "materials research laboratories" where university scientists, without breaking their disciplinary affiliations with the established departments in which materials had been traditionally investigated (geology for minerals, metallurgy for metals, and so on), could explore the theoretical leverage that quantum mechanics provided for the study of materials. Corporate and government laboratories increased their hiring of scientists trained in atomic theory to support their investigations of materials that interested them. The recent desire for university research in technologically interesting materials has led to collaborations rather than reformed materials research laboratories because the universities do not wish to disrupt well-functioning laboratories and the funding agencies do not wish to support university scientists to investigate such materials in isolation from non-university laboratories. Collaborations in materials science are thus multi-sectoral.

The Science and Technology Center for High-Temperature Superconductivity consists of the University of Illinois, the University of Chicago, Northwestern University, and Argonne National Laboratory. The two universities had NSF-funded materials research laboratories, but the juxtaposition of the discovery of high-temperature superconductors with NSF's decision to create a new program office, the Office of Science and Technology Infrastructure, inspired numerous researchers to wonder whether their research skills would be better supported and more valuable if they were part of a superconductivity center. Pairing themselves with a government rather than a corporate laboratory seemed sensible in view of the lack of immediate industrial applications for high-temperature superconductors, and Argonne had research facilities for materials research. The geographic proximity of the four organizations promised to facilitate the circulation of people and samples among the sites. The organizations' administrators organized workshops of the potential participants, and by sifting, combining, and aggregating the ongoing and potential research programs of these scientists, the administrators hoped to build a proposal that covered the major research challenges to understanding and using high-temperature superconductors.

Various organizational needs followed from drafting a multi-organizational proposal by reconfiguring the research plans of individual scientists from several organizations and multiple sectors. The collaboration needed rules for managing money among the organizations because research interests did not cleanly coincide with organizational boundaries and because NSF did not want its funds to support Argonne scientists that the Department of Energy was already committed to supporting. The organizations' administrators negotiated a formal contract that specified how funds should be divided among them and set the terms of Argonne's participation.[35] Making the collaboration's research plans fit within the limits set by the contract became an ongoing exercise that made "science representatives" highly influential in collaboration affairs:

We have middle managers. Now, there is a kind of tension. We have a matrix management system. The columns are the sites [meaning organizations], and the rows are the disciplines. Right now we are divided into three scientific themes and two smaller projects, and there's a little bit of mixing up between them. So the theme and project leaders have authority to meet with their people [regardless of the organization they work for] and put together the strongest possible program. . . . These are the people I really try to listen to as much as I can. . . . It comes back to the vision and the responsiveness of these middle management people to the needs of the field and how the center can meet these needs.

The power of these "science representatives" to shape the research directions of this collaboration far exceeds what group leaders in particle-physics collaborations can exercise (because much is decided in collaboration-wide meetings in particle-physics collaborations) or what the organizational leaders of BIMA can exercise (because the member organizations of BIMA create panels to review the proposals of its individual scientists and to allocate observing time).

Simultaneously, forming a collaboration by reconfiguring individuals research programs freed the collaboration from acquiring and integrating large amounts of instrumentation. The strength of the collaboration's proposal was based on continuity between its participants' proposed and previous research, so the individual participants already had functioning instrumentation. For example, well before the discovery of high-temperature superconductors, a Northwestern University chemist began investigating the synthesis and characteristics of complex oxides of copper, manganese, and other elements:

> In my mind there were two groups in the world working on that question. . . . It was a pretty uninteresting subject, I guess. There certainly wasn't a lot of wide interest in it around the world. . . . And so when the IBM Zurich people then published . . . that paper when they discovered [high-temperature] superconductivity, . . . as soon as we read the paper we exactly knew the chemistry, we knew what compounds to make and how to make them too, *and we had the facilities.* . . . We were the first to really document that X had in fact made a superconductor above 90 degrees. . . . We were working in this field, but not from the point of view necessarily of superconductivity, but from a point of view . . . of how to make them, what their structures were, and as a chemist we typically didn't get so much into the properties, but we were always interested in that.

While this investigator improved his instrumentation with collaboration funding, he did not undertake anything like the design and construction of new transition radiation detectors, as did FNAL 715, or the acquisition of digital seismometers, as did IRIS. The other participants in the Superconductivity Center did not impose deadlines on him because their work was not hostage to his acquisition of improved instrumentation. For him (and the other participants), joining the Superconductivity Center was a matter of using his research skills and his laboratory's instrumentation primarily to further understanding of high-temperature superconductors rather than primarily to advance the art of chemical synthesis (or other arts peculiar to the other disciplines represented in the Center).

Sectoral Instigation
Universities were the most important sector in many collaborations, and they were almost never insignificant or absent in the collaborations examined here.[36] In terms of their general practices and the consequences of their inclusion, differences among universities seem less important than commonalities. Hence, the primary distinction in the collaborations we studied was between those that were instigated exclusively by universities and all others. We expected collaborations with multi-sectoral instigation to face greater formative complexities than university-instigated collaborations. We expected collaborations with corporate instigators to be encumbered in their formation because of corporate needs to secure intellectual property. Figure 1.6 illustrates the relationships of sectoral instigation with five other dimensions. Collaborations with multi-sectoral (or non-university) instigation were usually larger (defined as seven or more organizations or teams), more likely to have a communications center, more likely to have an agreement about sharing data, but also less likely to go over budget.

The source of these relationships is the need for instigators outside the university context to make a collaboration "university-friendly," even when a non-university sector dominated the collaboration. Once again, consider the Active Magnetospheric Particle Tracer Experiment, whose formation was dominated *not* by universities but by research institutes that could build functionally differentiated spacecraft for the release and detection of tracer ions. However, the Applied Physics Laboratory, the Max Planck Institute for Extra-Terrestrial Physics, and the Rutherford-Appleton Laboratory were not viewed as preeminent in all the instrumentation desirable for the spacecraft. The more uses to which the spacecraft could be put, the better the project would look to NASA, its advisory committees of external scientists, and science policy makers in Germany and the United Kingdom. Consequently, the instigators sought universities and corporate groups with relevant expertise and as a result the number of participating teams and organizations ballooned.

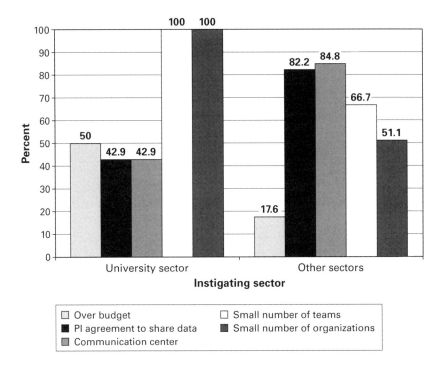

Figure 1.6
Sectoral homogeneity by size (teams, organizations), communication center, principal investigator's agreement to share data, and cost effectiveness.

A rough quid pro quo governed the inclusion of the many teams. The university and corporate teams made two major concessions to the research institutes that instigated the collaboration. First, the teams granted the institutes, and particularly their engineers in charge of spacecraft design and construction, a role as communications centers with responsibility for using the information provided by the teams to identify and solve all manner of interface problems. The scientists who led teams from outside the research institutes understood they were to resolve their conflicts with these engineers directly, and that the engineers had the confidence of the instigating scientists.

Scientist: To get the resources, this was ultimately negotiating, in this case, with APL [Applied Physics Laboratory]. If you came to a point where you said "Absolutely need additional weight," you make your case strong enough. . . .

Interviewer: To whom are you making your case: Krimiges [the scientific leader] or Dassoulas [the engineering leader]?

62   Chapter 1

Scientist: It really depends. . . . Early on, of course, it's Krimiges. Before you get to detailed design. By the time you are talking about detailed interfaces with spacecraft, then, at that point—weight, power, etc.—it was with Dassoulas.

These engineers, as a matter of both training and experience, knew how to track budgets and how much things cost. Their empowerment over the interfaces between scientific instruments and the rest of the spacecraft led to the design and construction being finished within budget. The second concession of the teams was to share their data in a collaboration-wide database during the time the active experiment was being run—when the spacecraft outside the magnetosphere released tracer ions for the spacecraft inside the magnetosphere to detect. In return, university scientists received two privileges. First, they became recognized secondary participants in the active experiment (all understood that Krimiges was the author of the active experiment). Second, and more important, they had the autonomous use of their instruments to pursue their individual interests when the active experiment was not being run. As the results of the active experiment were disappointing, the individual teams' measurements of the natural environment assumed greater importance (Krimiges et al. 1986).

Organizational Context
In about 40 percent of the cases we studied the individual scientists who wished to participate in multi-organizational collaborations needed to secure approvals from parent organizations or experienced similar constraints. Collaborations that faced this aspect of formative complexity usually had large numbers of teams that did not communicate often,[37] formal contracts, a clearly established system of rules, and many levels of authority (figure 1.7). These features are to be expected when organizational members are concerned with how the collaboration will affect their reputation and ongoing interests.

The history of the Incorporated Research Institutes for Seismology embodies this dynamic. A few universities had begun investigating means for acquiring large data sets to address issues in academic seismology. Seismologists at those universities enjoyed superior computation facilities, research-and-development laboratories, or good connections to industry or government agencies. IRIS's proponents among these fortunate pioneers assumed that IRIS had to be large to impress NSF with the breadth of support for IRIS. Moreover, a majority of members could inhibit the fortunate minority from continuing to compete when coordination better served the seismological community's interests. In order to get numerous

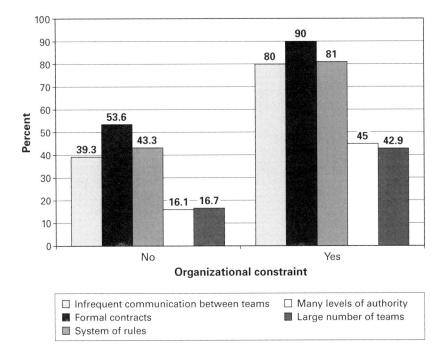

Figure 1.7
Organizational constraint by size (teams), levels of authority, system of rules, formal contracts, and frequency of communication between teams.

universities without seismological facilities to join IRIS and to put up some money for start-up costs, IRIS had to develop an elaborate formal structure. Otherwise, the seismologists at the non-pioneering universities would have trouble convincing their administrators that IRIS would be fairly and equitably governed.

The formal structure included the specification of the three distinct programs that IRIS would perform: build up a digital global seismic network, build up a stable of digital portable seismometers, and build up a data-management center to support the handling of large data sets. Each program would have its own budget, its own officer in charge of implementation, and its own advisory panel of seismologists from IRIS organizations to set goals and review progress. The three officers reported to a consortium president who was accountable to the board of directors, which had an active executive committee to monitor collaboration-wide issues. No elaborate inter-program technical integration was needed for each officer and advisory panel to proceed with its work. The member organizations had only to communicate at annual meetings of the board of directors, at which the consortium's budget was set. Initially, this communication was more meaningful than anyone anticipated, because the directors made the IRIS budget an occasion to discuss the possibility of redressing historical inequities in the distribution of resources among universities:

> When those things [specifying IRIS's functions] started surfacing, they didn't sit too well with some folks who had something to lose, some autonomy to lose. . . . Some of the people who started the whole thing had these big, eclectic meetings, with all the seismologists . . . where this was laid out as they were going to save every little university, and "You've got to get on board because there's no self-serving thing here, this is all for you. We need your backing." Then, suddenly, it turned out, "Well, it was all for you, but it was all for me, too." . . . There was a subset of membership that didn't want to spend this [IRIS] money on [a data-management center]. . . . Those who didn't need it, and in fact would benefit without it, because it would keep their position relatively high in such facilities, and it would put more IRIS money into things that they couldn't provide for themselves. So, it was a fight every year to distribute the budget, and make sure that DMC got something. It was the ugly aunt that we had to keep dragging out of the closet and say "She's going to eat with us."

When collaboration requires that the interested scientists secure the approval of their home organizations, a collaboration can easily become an arena in which broader conflicts among member organizations are played out.

Conclusion

That multi-organizational collaborations form in different ways comes as no great surprise. The multiplicity of funding sources in the United States, the complexities of combining support from multiple governments in international collaborations or from multiple agencies in the same national government, and the range of scientific specialties that spawn collaborations all augur for diversity in the formation of collaborations. What is significant is that patterns in the formation of multi-organizational collaborations do not coincide with scientific specialty or patronage. Particle-physics collaborations are the exception, but even these are not uniform. For every variety of formative process, we could choose an illustrative case from a broad array of specialties and patrons. Though our data set is weighted toward American collaborations, two of our illustrative cases, AMPTE and FNAL 715, were international. Still, further research is desirable to ascertain the extent to which idiosyncrasies of U.S. science policy is responsible for our findings.

At least in terms of the ways they form collaborations, scientific specialties exhibit internal heterogeneity while their patrons exhibit flexibility. The contexts that give rise to multi-organizational collaborations have much less to do with disciplinary tradition or long-standing legal frameworks than with research conditions that are important to most specialties and organizational developments that are important to most patrons. A nifty idea for measuring a parameter serves as a wake-up call to scientists, as will a new funding agency program that invites scientists to try new approaches to making measurements. Envy of the facilities enjoyed by scientists in government, industry, or other nations energizes university scientists to brave the politics of forming a large collaboration that could obtain such instrumentation. Aging facilities that cost too much for an organization to recapitalize or other sources of long-term professional insecurity inspires scientists to become entrepreneurial. Inadequate human or technical resources for addressing a scientific question motivates scientists to advertise for collaborators. Stellar research opportunities can lead agencies and program managers to organize impersonal competitions and eventually collaborations.

The relationships between project formation and other facets of multi-organizational collaborations demonstrate the utility and limits of research into origins. Our general argument is that the formative encumbrances of collaborations—that is, the complexities imposed at the outset—affect their structure. When a central authority pressures scientists to collaborate, when funding is from uncertain sources, when funding is from a reorganized source, or when the home organizations of individual participants weighed in on the collaboration's viability and plans, participants should expect relatively high organizational formality. Pressure from a central authority, the involvement of parent organizations, or the pursuit of novel funding sources usually forces participants into relationships with scientists and organizations they have no grounds for trusting. In such cases, collaborations can formalize their management and centralize their organization to ensure the appearance of fairness and accountability in decision making—or collaborations may limit their scope to ensure autonomy for the member organizations and their individual scientists.

Collaborations that form with support from a reorganized funding agency often reorient the research interests of their members. In such cases, the movement to collaborate is not only a choice to do something; it also involves opportunity costs—choices *not* to engage in alternative activities. The typical response of instigators faced with the prospect of dis-embedding members from ongoing projects is to create several levels of authority within the collaboration. Middle managers become important and powerful arbiters between the potentially conflicting pull of the participants' scientific interests and organizational loyalties. When organizations other than universities are significantly involved, professional engineers from the corporate or governmental world will likely be managers, but the participation of universities ensures that the collaboration accommodates academic sensibilities.

2
Magnitude

The term "Big Science" does not discriminate among big, bigger, and biggest or among ways of measuring size. Size may be indicated by number of scientists, total size of staff, amounts of money, length of time, or scale of instrumentation. The difficulties are also practical. To raise funds and recruit others, instigators of multi-organizational collaborations need to get the magnitude right: "Given what we want to do, how many organizations should be involved?" They want to know what their demands will be. Will they be subdivided into teams? How much is this going to cost? Can it be done for less, or should we ask for more? How many scientists, engineers, researchers, graduate students, administrative staff are needed? What kinds of resources will be requested by a new organization? How long is this project likely to take? How long until it can start? Should we start by inviting everyone who does work in this area and pare down the project later? Or should we work mainly with those we have worked with before?

Potential collaborators ask such questions because they sense that the answers will have consequences for the project's evolution and success. Organizational theorists are not alone in hypothesizing a relationship between size and other organizational features. Common sense generalizes anecdotal experience to tell us that larger organizations have certain characteristics that smaller ones do not: *other things equal*, larger collaborations are more bureaucratic than smaller ones.[1] Size, when it is not mere aggregation but accompanies greater resources and complexity, *should* be associated with greater formalization (Shrum and Morris 1990). Indeed, it is sometimes not "size" that really matters but its real and imagined consequences. Those who lamented the rise of "Big Science" in the 1960s were not worried about how many people were doing science. They fretted that the increase in the number of people needed to carry out a single research project led to a stultifying division of labor, reliance on professional managers with a different ethos from scientists, diffuse responsibility for tasks, and diluted credit for discoveries. That is, they were worried about bureaucracy—a form of social organization whose characteristics they feared were fundamentally incompatible with scientific research—and the amount of bureaucratically organized research that academic scientists could be tempted or compelled to pursue.

Such fears have influenced government policy in the United States. It is often hard to make a case for a gargantuan project undertaken for a significant period of time against the backdrop of resistance from researchers arguing that the amount of resources absorbed by "Big Science" leaves too little for "small science" endeavors that embody what is best in science. This reaction was, for example, one of the primary reasons for the failure of the Superconducting Super Collider project—particle physics was perceived by other physicists as gobbling up a disproportionately large amount of money relative to other fields of physics (Kevles 1997).[2]

The research structures that became prominent in subatomic physics after World War II—the national laboratories, their particle accelerators, nuclear reactors, research groups using bubble chambers—dwarfed even the university cyclotron laboratories of the 1930s, to say nothing of the 1930s laboratories that probed nuclear structure by making neutron beams with radium and paraffin. But the sizes of the postwar structures did not prevent particle physicists of the 1970s from introducing a bevy of electronic detectors that supplanted bubble chambers or from developing a new, philosophically interesting ontology of elementary particles and their relationships. Sciences with longer traditions of fostering large-scale structures—including astronomy (with its observatories) and oceanography (with its research vessels)—have not scaled back to remain vibrant. Today, even researchers in biology and materials science, which have little history of large-scale research, bring instrumentation and samples to national laboratories to perform experiments with the intense monochromatic radiation given off by electron accelerators. The scientific descendants of the particle physicists who first bemoaned "Big Science" do not fret over how working with larger particle accelerators will affect their pursuit of new knowledge but bemoan the short-sightedness of legislators who declined to see the Super Collider through to completion, and the selfishness of scientists who opposed it as a drain on public funds that could instead be spent on smaller projects. Either size has not entailed the levels of bureaucracy that the critics of "Big Science" feared or bureaucracy and scientific research are more compatible than they imagined.

Size, Magnitude, and Their Measures

For Knorr Cetina, the principal sociological question of particle-physics experiments today is posed by size: How is it possible to run an experiment with so many participants lasting so long (1999)? How shall we understand these enterprises intellectually? How will we organize them practically? Given that some research projects are always larger than others for a particular time, place, and specialty, do larger projects work differently from smaller?[3] Do larger collaborations create pressure to centralize authority for purposes of coordination or do they delegate authority to take best advantage of the strengths of their component parts?[4] What limits are there on the size of research collaborations?

Any effort to deal with such questions should delineate the various meanings and measures of "big." Size can refer to a variety of aspects in scientific research: the character of instrumentation; the scope of research; the number of people, organizations, or teams; the funds required, available, or utilized; the time required to get the project off the ground; its duration. These parameters are not equally easy to define or to measure, and they should not be assumed *a priori* to lead to the same relationships with other characteristics of multi-organizational collaborations.

"Magnitude" is preferable to size as a generalized concept because it conveys variations in numbers of actors, project duration, and cost. This chapter examines the first two of these parameters. Although we were initially diligent in seeking information on the costs of collaborations, the idiosyncrasies of particular specialties, differences in methods of accounting, proprietary information, and differences between organizational and collaborative resources rendered the issue of cost too difficult to develop a comparative indicator of magnitude.[5] Numbers of participants and duration are not themselves simple matters. The constituent "actors" in a collaboration may be people, teams, or organizations.[6] Duration, too, has its ambiguities. Does collaboration begin when an idea is formed? When a proposal is drafted? When money is available? When data are first taken? Does collaboration cease when data are no longer collected? When publications are no longer forthcoming? When funding is no longer available? Collaboration may have many "beginnings" and many "ends." We attempt to capture that complexity by measuring duration in multiple ways.

Focusing on indicators that could be generally understood by our informants, reliably estimated, and compared across projects, we asked about the numbers of participants, the numbers of teams, the numbers of organizations, the time needed to acquire funding for a collaboration, and the time needed to obtain scientific results once funding was obtained. Collaborations in the fields examined here are clearly larger than most scientific projects, but smaller than some of the recent particle-physics mega-experiments that involve up to 1,500 participants from 160 separate organizations, lasting up to 20 years (Knorr Cetina 1999). Collaborations are dynamic social formations. If successful, they usually grow—they will, at minimum, add graduate and postdoctoral students. They can also shrink, as members obtain positions at other organizations and are not replaced. Sometimes participants leave an organization and recruit a new organization and its facilities to expand on the original collaboration.

The average number of individual participants in these collaborations was more than 50, ranging from 8 (SLAC 137) to 150 (IRIS). Not included as individual participants were people who would have performed the same task had the collaboration not been conducting or supporting research. For example, we did not include as collaborators in space-science projects the people who launched the satellite: they perform that task whether a satellite is for research, communication, or intelligence gathering. For crews that sail oceanographic research vessels, the same consideration applies. On the other hand, we did include as collaborators people who helped to design and operate research instrumentation regardless of whether they were designated scientists, engineers, or technicians.

The number of organizations ranged from 3 (BIMA, CPIMA, CTA-CTP) to 35 (IRIS at the time we conducted interviews), averaging more than 8 per collaboration. Like the number of individual participants the number of organizational participants can vary during the course of collaboration. Additions are more common than losses, but there were occasional instances of organizations leaving collaborations. When collaborations expanded because their initial success attracted new members, we used the number of organizations responsible for the initial success. When collaborations expanded because the ambitions of the original members expanded and were no longer credible unless they attracted new organizations, we used the number of organizations in the enlarged collaboration. When collaborations contracted because some of their organizations thought the resources they helped create were no longer serving their interests, we used the number of organizations responsible for creating the resources. The number of participating organizations is related to the sectoral origin and composition of collaboration. For many scientific projects, available expertise is located predominantly in universities. Collaborations involving only the university sector typically involved *fewer* participating organizations and teams. More than 90 percent (91.7 percent) of such projects involved only 3–6 organizations, versus 47.5 percent of collaborations with participants from multiple sectors.

Strictly speaking, science is performed not by organizations but by research teams. In many collaborations, teams and organizations coincided, but occasionally, research teams within a collaboration were themselves multi-organizational. In some cases, teams were defined topically in ways that transcended the boundaries of the member organizations. In our sample of collaborations the average number of research teams is slightly higher than the number of organizations involved.[7]

What is most important in understanding the question of magnitude is that "project requirements" are not an *a priori* force that determines project size. A multitude of considerations result in decisions to define project requirements one way rather than another, the collaborators themselves are not always the decision makers, and the position of the decision makers can vary with scientific specialty. For example, accelerator laboratory directors and their advisory committees were more important powers in determining the size of particle-physics collaborations than program managers at funding agencies. The opposite was true for directors of space-flight centers and NASA program managers in space-science collaborations.

In making these determinations there are pressures in both directions: to limit and to increase participation. Pressure to increase participation results from an interest in generating political support, while pressure to limit participation results from fears of conflict and bureaucracy. In some cases, funding exigencies seem to demand an increase in the size of a project:

> Cynically, one could suggest that several of the universities were brought in to show a large university presence during the proposal phase. . . . We felt a little bit like . . . a car mechanic as part of a racing team. We weren't going to drive, we weren't going to make decisions about which car to use, or even what races to participate in but we would work on the car. . . . They were really telling us "Well we needed your name to help us get this thing, but now that we've got it . . . there's really not much of a role for you."

Although the university participants in this project were not pleased with their level of involvement, their participation may have been crucial for the collaboration as a whole—a racing team needs mechanics. Alternatively, as we saw in the last chapter, the pre-existing structure of relationships can be crucial:

> We were concerned with not making the group too large, since we knew that it could become unwieldy. Because the co-investigators knew each other and in many cases had collaborated . . . we knew that there was a good basis of collegiality and interaction that would allow us to succeed as a group. So I don't believe that anyone else was asked.

In this case, pre-existing relations are not the core of a larger group, but serve as a kind of natural boundary to the size of the collaboration.

The second component of magnitude is duration, the temporal dimension that distinguishes short projects from lengthy ones. Though actual termination cannot be accurately predicted, the fact that collaboration is generally *designed* to end compels us to consider length as an aspect of magnitude. Yet "lifetime" measures of collaborations are difficult for some projects. Papers continue to be published after funding runs out. Structures remain in place pending decisions on a new grant proposal requesting support for follow-up studies. Lifetime measures simply aren't applicable to many of the collaborations we examined, so we settled on two alternative indicators of duration: from inception to funding and from funding to first publication. Time to funding was the most significant aspect of magnitude, since the amount of effort required to form a collaboration has consequences for its subsequent operation.[8] Although it is possible for a collaboration to be short in duration but large in number of participants, this tends not to be the case. Collaborations that take a long time to acquire funding or to generate scientific results are generally seeking to mobilize many participants.[9]

Duration to funding[10] establishes a scale for the project's ambition and its needs relative to the particular field of science and to the potential funding sources.[11] "Initial formulation" of the idea for a collaboration refers to the year of the instigator's first attempts to convince colleagues and patrons to form the collaboration and not to earlier failed attempts at forming collaborations that might have been somewhat similar. Robert Wilson, for instance, twice failed in the early and mid 1960s to convince ESRO/ESA to build and launch a satellite for ultraviolet astronomy into low Earth orbit. We did not view those proposals as the initial formulation of what became the International Ultraviolet Explorer. Instead we used

1968—the year Wilson made connections with NASA's Goddard Space Flight Center, whose officials urged redesigning the satellite for launch on the next generation of American rockets into geo-synchronous orbit, where it could be far more easily and productively operated.

In contrast, for years scientists at the Applied Physics Laboratory and scientists at the Max Planck Institute for Extra-Terrestrial Physics together nursed the idea of an Active Magnetospheric Particle Tracer Experiment on minimal grants and internal funds until an opportunity arose for them to compete for funding for detailed design studies and construction. The initial formulation was the recognition by scientists at the two laboratories of their common interests and complimentary skills, even though they built a more sophisticated set of spacecraft in the 1980s than they were able to envision early in the 1970s.

The average length of time from the initial formulation of a project to the point of funding was 2¼ years. Minimally, collaborations require less than 6 months before resources are acquired (VLBI Network). The longest wait in our sample was 9 years (Hobby-Eberly Telescope). More than half of the collaborations in our sample waited 21 months or longer. Duration to publication indicates the amount of effort required before the project began to yield scientific results. In many instances, collaboration members published papers on techniques or instrumentation they were developing for the collaboration. Because interviewees unambiguously distinguished these papers from published scientific results, we did not consider these publications "scientific." Thus, the greater the R&D work needed to obtain data (relative to collaborative resources), the greater the duration to first publication of scientific results. These projects took an average of 3½ years after the initial idea before publication, although one collaboration (the Grand Challenge Cosmology Consortium, which generated "data" by simulating cosmological processes and thus did not have to build or purchase instrumentation) began publishing after only a few months. In contrast, the Einstein Observatory and the Active Magnetospheric Particle Tracer Experiment did not begin to publish results until 9 years after inception. A comparison of the duration to funding and publication shows that slightly more than a year elapses between funding and publication: collaborations have already been germinating for more than 2 years.

Our analysis indicates that larger collaborations are indeed more elaborate, but not necessarily more hierarchical in decision making about matters most directly bearing on the generation of scientific results. As a rule, collaborations either keep deliberations over such matters broadly participatory or empower their individual researchers, teams, or organizations to make the necessary decisions. Either way, the scientists come away feeling that the size of the collaboration, even when extraordinary, did not foster social relations that undermined what they consider quintessential to science. A relationship between magnitude and decision making exists only for decisions about money, time, or other resources, and the relationship is not straightforwardly linear across all the variables that can be used to indicate size. Depending on which form of magnitude we mean, different forms of decision making become prevalent as collaborations become larger. As the number of participants increases, an oligarchy emerges to make decisions about resources. As the duration increases, collaborations tend to involve a corporate board of nonparticipants.

Magnitude and Formalization

The magnitude of inter-organization collaborations is positively related to the degree of formal organization and management (figures 2.1 and 2.2). Larger collaborations—whether viewed in terms of number of participants, organizations, or teams—are more likely to have formal contracts, more levels of authority, systems of rules, and administrative leaders. These factors do not themselves imply centralization of authority, but they do imply a need for explicitly articulating the activities undertaken by the collaboration and a need for maintaining confidence in the mutual fulfillment of commitments.

Large, formally organized collaborations are especially prevalent in the field sciences (e.g., space science, geophysics); small, informally organized collaborations are more common among laboratory sciences (e.g., materials science). In general, laboratory scientists have more flexibility to oscillate between taking data and improving their instrumentation, because they are themselves creating the conditions under which data are collected. Laboratory scientists use that flexibility to create relatively small, informal collaborations that fluidly reconfigure themselves in response to shifts in the interests of their members and the broader community.[12] Field scientists, in contrast, must take advantage of natural events when they occur and wait for politically and logistically appropriate times to mount expeditions to inhospitable locales. Because opportunities to form field-science collaborations have been less frequent and less regular, they have interested broad swathes of the relevant disciplinary communities and required proponents to include as many partici-

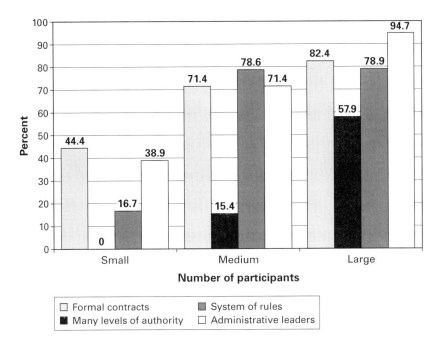

Figure 2.1
Number of participants by formal contracts, levels of authority, system of rules, and administrative leaders.

pants as were fiscally and technically feasible.[13] With more limited chances to collect data at the same time and place as their colleagues, field scientists usually segregate their efforts to develop innovative, technically risky instrumentation from projects that place instruments in scientifically attractive but hostile locales.[14]

The Active Magnetospheric Particle Tracer Experiment (AMPTE), our example of a conventional collaboration with a dominant sector, illustrates the positive relationship between magnitude and organizational formalization. The collaboration began as an intriguing idea for a tracer experiment to evaluate theories on the origins of magnetospheric ions, but to become a proposal that could win a NASA competition, it mushroomed into a tracer experiment plus a bevy of less risky, complementary objectives. The latter were, occasionally, more important to some members of the collaboration than the tracer experiment. Even after winning approval from NASA the leading scientists in the collaboration pushed for expanding the collaboration to make use of the maximum weight that NASA's launch vehicle could handle.[15] In all, AMPTE

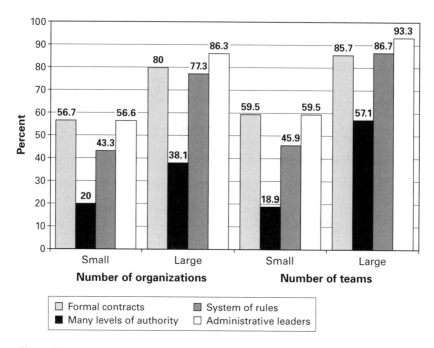

Figure 2.2
Number of organizations and number of teams by formal contracts, levels of authority, system of rules, and administrative leaders.

combined the efforts of three nations, 130 scientists, engineers, postdocs, and graduate students, and 18 organizations to build three spacecraft that were launched together and deployed into different regions of space.

To maintain accountability to these governments, the collaboration formalized its relations through numerous contracts and subcontracts. At the top of the hierarchy was an international agreement between NASA and the German Ministry for Aeronautics and Astronautics. NASA's Goddard Space Flight Center issued a contract for the American spacecraft to the Applied Physics Laboratory, which in turn issued subcontracts to American organizations that designed and built scientific instruments. The German Ministry funded the Max Planck Institute for Extra-Terrestrial Physics, which built the German spacecraft and arranged for the contributions of other German organizations. The British scientists joined AMPTE through a subsidiary contract between the German Ministry and the United Kingdom's Science and Engineering Research Council, which funded the Rutherford-Appleton Laboratory to build the U.K. portion of the spacecraft.

The elaborate set of contracts and subcontracts was matched by an elaborate organization with multiple communication lines to the authorities at the funding agencies. At the international level, an experienced Goddard project manager insisted on regular collaboration-wide meetings for ensuring the successful integration and launch of the three spacecraft:

> It was my management technique, and I've used it on a bunch of projects, where you start the meeting the first day, and maybe there's 150 people there, and you get everybody together and tell them for two hours what's happened since the last meeting. . . . Then you split them up into working groups where the scientists sit in one place, spacecraft people in another, and launch vehicle people in another, ground data system people in another. And they meet in parallel sessions and talk about their particular items, and then I get two groups together and have them talk to each other, to select which groups, and then, finally, we all get back together, and I have each one of those groups stand up and report what happened with the major thrust was, what the major action items were, what they expect to come out of it before the next meeting.

For each participating nation, a designated manager was responsible for defining and overseeing the technical interfaces among the spacecraft systems, including their suites of scientific instruments, coordinating the construction of the spacecraft, and controlling the budget.[16] These managers were professionals with positions that were independent of their management of any particular project. Their responsibilities made them powerful in their dealings with scientists contributing instrumentation, who often competed for the resources controlled by the managers. However, the managers were obliged formally to consider the three scientific leaders as more-than-equals (though not quite superiors). The scientific leaders of AMPTE had their own lines of communication to their funding agencies' headquarters. Each manager had a supervisor, and the scientific leaders were expected to inform headquarters if they felt that managerial decisions were unreasonably compromising the scientific capabilities of the project.

The contrasting case of a small and non-bureaucratic collaboration is exemplified by one of the ground-based astronomy projects that coordinated several extant radio observatories to perform very-long-baseline interferometry. The Sagittarius A project involved fewer than 20 participants, who formed six teams, each of which operated a radio observatory.[17] They observed the radio source Sagittarius A at a wavelength of 3.5 millimeters, an order of magnitude lower than what was customarily used in VLBI observations circa 1980.[18]

The informal organization and management of the collaboration was predicated on two factors. First, only a small number of radio astronomers had the technical sophistication and ambition to attempt such an observation and these had a history of working together to observe objects of shared interest or to stretch the state of the art in VLBI. As a result of their shared experience in making successful VLBI observations,

> there's a certain amount of well-established arrangement in that respect [the organization for making observations], so it relies on basically people doing their best. But there is no formal arrangement.

Second, the collaboration did not need explicitly dedicated funding to attempt the observation:

> Even though individual observatories might have been supported by the National Science Foundation, this was not an experiment that was, as an experiment, really supported by the Science Foundation. It's just that the individual observatories were. We [the senior members of the collaboration] proposed separately to each one of these observatories and got the [observing] time.

In the absence of a need for financial accountability, no contracts were drafted to establish responsibilities and reporting requirements. Senior participants understood each other's roles and relied on their in-house discretionary funds for their needs. No formal managerial positions were required to judge jurisdictional conflicts or to divide resources.

Through the use of e-mail, the collaboration functioned with an informal hierarchy that recognized the individual scientist most responsible for convincing others to attempt the observation. That individual led the electronic discussion of the observing strategy, made sure that the observatories scheduled the correct observing times and had the necessary equipment, and made on-the-spot judgments as needed during the actual observations (e.g., what to do if one of the observatories was having weather or instrumentation problems). Comparing the hierarchy of authority to a university department, one participant said:

> I think there were fewer [levels]. I think there's probably only one level in this case, and that's the PI, or the guy who is actually running the experiment, and then everybody else. I think it's essentially a two-tier structure.

Such a structure worked perfectly well when a project could be handled by a half dozen experienced scientists with similar skills and mutual respect—plus students, postdocs, and other immediate associates.

Magnitude and Decision Making

Larger projects may be more formal, but this does not necessarily imply hierarchical decision making. The magnitude of collaboration was *not* significantly related to the actors that made decisions on scientific or personnel issues. This null result deserves emphasis because of the propensity to assume that larger collaborations are bureaucratic *and* centralized in decision making. Size does lead to more bureaucracy, but bureaucracy need not be oppressive, and it may be confined to areas where it does the most good and least harm. For example, it is standard procedure in particle-physics collaborations, whether large or small, for individual participants to submit their substantive scientific claims to collective review and approval before any manuscript is submitted to a journal. In large collaborations, this review is often more bureaucratic because large collaborations frequently create a committee to clear drafts of potential articles for collaboration-wide review. But the principle of vesting decision-making power in the whole collaboration remained the same. As we will see in the next chapter, descriptions of these intra-collaboration reviews ranged from useful to intellectually exhilarating. Virtually no one described them as oppressive, petty, or an obtuse elevation of process over substance.[19]

Or consider ground-based astronomy collaborations that built observatories. Whether large or small, they could be considered bureaucratic because they operated by letting contracts for the construction of the several observatory subsystems. The larger ones may be deemed more bureaucratic because they subjected designs to external reviews before construction contracts were let. But regardless of the collaboration's magnitude and the presence of external review committees, participating university scientists generally found these operations oppressive because of the restrictions they imposed on their ability to revisit design questions and try out new ideas. Such operations usually excluded graduate students from participation in this aspect of observatory development.[20]

Both the temporal and the participation measures of magnitude are meaningful in decision making about resource issues: the distribution or use of money and occasionally the use of time or access to instrumentation provided to members. Smaller and shorter collaborations are more likely to make decisions about resource issues in a participatory manner or to empower a scientific leader to make such decisions. They are less likely to empower a corporate board, an oligarchy of participants or an administrative leader to make decisions. However, as collaborations became larger in size and needed longer periods to acquire funding, they did not uniformly empower boards, oligarchies, or administrative leaders. Oligarchies were most important in larger collaborations, and administrators were most important in longer collaborations. But the relationship is not simple. In terms of duration, oligarchies peaked in importance in the medium-term projects—not the longest ones. In terms of size, corporate boards peaked in importance in projects with moderate numbers of participants. These differences are due to divergences in the reasons some collaborations are large and the reasons some collaborations take longer to fund.[21]

Fermilab 715, which was used to illustrate the "collaborators wanted" form of project formation, also serves as an example of a rapidly initiated collaboration that did not become hierarchical in decision making while growing to moderate size. The collaboration grew from its core of Yale and Fermilab physicists to add capabilities in electron detection and to show that it could draw in its competitors. But except for the transition radiation detectors, built and maintained entirely with Soviet resources, the experiment used reconfigured and updated instrumentation that the American participants, in many cases, already knew intimately. Thus, the American physicists did not need to draft a proposal for funding agencies to consider, and thus they did not need an authoritative decision maker to allocate money or engineering resources. The collaboration did need to decide how best to use the limited beam time it received from the accelerator laboratory. But participants considered use of beam time a quintessentially "physics" issue to be decided by consensus reached in collaboration-wide discussions rather than an "engineering" issue to be vested in a decision maker who could be held accountable for technical incompetence or fiscal dishonesty.

Of the 22 collaborations that were funded or initiated quickly, only five built substantial instrumentation from scratch. Collaborations working with proven, extant instrumentation do not impose much hierarchy over resource-related decisions, because few fear needing to shift resources to deal with unanticipated difficulties or unanticipated successes. Either the collaboration as a whole can be trusted not to become bogged down over a small resource issue or a scientific leader can be trusted to be fair when the stakes over resource issues were low. As the time it took to acquire funding increased, corporate bodies and administrative leaders became the most important decision makers over resources. As the number of participants increased, oligarchic decision-making bodies became most important. Keck and AMPTE respectively illustrate these processes.

The Keck Observatory was a collaboration of several University of California campuses and the California Institute of Technology, with the University of Hawaii and NASA in lesser roles. It represents those collaborations that take a great deal of time to fund, with a corporate board and administrative leader dominating decision making about resources. Keck's origins lay in the diminishing attractiveness of the University of California's Lick Observatory in the late 1970s.[22] Enthusiasm was generated when an astronomer and a physicist came up with competing ideas for building a far larger, more powerful telescope rather than making an incremental improvement on the latest Lick telescope.[23] Developing ideas for telescopes of unprecedented size takes time, as does putting together a blue-ribbon panel that would be perceived by all as capable of rendering a fair and definitive judgment on the merits of the competing ideas. Furthermore, a new University of California observatory was not something any combination of UC scientists could petition the federal government to fund. The UC system had to find patrons. University administrators did not begin raising money for a new observatory until they could present prospective donors with a thoroughly vetted plan that the astronomers were united in backing.

Two to three years after astronomers began to discuss the potential of a huge telescope, the university system began raising funds on behalf of the campuses for a new observatory with a telescope that followed the physicist's design. The university's campaign failed in the early 1980s until a single donor came forward to offer the bulk of the cost of building the observatory. The prospect of the gift led UC to ask the California Institute of Technology (Caltech) for the balance of needed capital in exchange for a proportionate share of the observing time. Before Caltech could begin raising funds, Howard Keck came forward to offer the money to build an entire observatory. Briefly, it appeared that the universities had the funds for two telescopes, but because of the donor's death and legal disputes the gift to UC failed to materialize. Caltech and the UC system agreed to build an observatory as equal partners, with Caltech providing the capital and UC providing the operating funds, the telescope's design, and the telescope's designers.

Considerable time was required for funding of the Keck Observatory because its ambitions were enormous (and thus had to be detailed and vetted), its membership and intended users were few (and thus made the project ineligible for federal funding), and it did not attract a solid foundation of donors from whom ongoing support could be expected but rather a single patron who wished to be equated with the observatory.

Under these conditions, no bench scientist or committee of bench scientists inexperienced in the intricacies of project management could be entrusted with the "nest egg" and the negotiations with suppliers and construction firms. To build the observatory, the universities' administrators created the California Association for Research in Astronomy (CARA). This body consisted of a board of trustees with equal representation from each organization. A project manager was hired for daily administration of the project. The Caltech members of the board, whose organization managed the Jet Propulsion Laboratory, saw to it that the manager was a hard-nosed, scrupulously by-the-book veteran of the Jet Propulsion Laboratory's spacecraft-construction projects.[24]

Astronomers on the Caltech and UC campuses and the physicists who conceived the mirror's design formed a Science Steering Committee. Yet this committee did not do much steering outside the realm of designing and building a suite of scientific instruments to be placed at the telescope's focal point. The committee that mattered most was the board of trustees, which, in the view of the Science Steering Committee chairperson, had "played an incredible role":

> [The board of trustees] met once every two months during this period [observatory construction] . . . a phenomenal interval [compared to] an academic organization in which the board of trustees might normally meet once a year. . . . But they were hands-on managers. They were paying attention.

Astronomers and physicists continued to debate and refine their ideas on telescope design as CARA began letting contracts for construction of observatory subsystems. But when they sought to steer the project, they ran into a brick wall:

> [I] would say in most practical instances the project manager outranked the project scientist. His job was building the telescope on schedule and on budget, and so he decided how the money was spent and how the CARA staff would spend its time. His word was the ultimate law. And I had to worry about whether we were building the right thing, so I worried about lots of technical details and had many arguments and debates with the project manager about the direction we were going in. . . . And we in fact came into conflict many, many times.

Occasionally, the scientists would seek to appeal over the head of the project manager to the board of trustees, but usually with little effect:

> The bottom line is that [the project manager] is a very strong guy. I told you that the board was important, . . . [and] within the board itself, the astronomers were not very important. It was really the officials from the upper echelons of the university. . . . They appointed [him] as their designated individual. [X], as a result,

had the ultimate say—[X] and the board together. . . . The astronomers would make requests, more than requests, and they would not be honored and in general, there was an atmosphere of tension there.

None of this should be construed to mean that the astronomers did not admire, respect, and come to appreciate the project manager. The telescope, with its segmented, spherically symmetric 10-meter mirror, was completed on time and on budget. And the astronomers realized in retrospect that empowering the board and the project manager was a beneficial idea, since the Science Steering Committee was unable to complete a scientific instrument on time and on budget.

AMPTE grew large in terms of participants and teams and established an oligarchy of scientists as the decision makers on resources. As described above, AMPTE formed at the intersection of Tom Krimiges's idea for a tracer experiment to investigate the source of charged particles in the Earth's radiation belt and the complementary technical specializations of the Applied Physics Laboratory and the Max Planck Institute for Extra-Terrestrial Physics (MPIET). The idea was risky and could yield a null result. To make it worth the MPIET's effort to design and build a suitable ion-release spacecraft, there had to be an opportunity to conduct ion releases their scientists could observe from the ground. To justify NASA's support, there had to be an opportunity for American scientists interested in the regions through which the AMPTE spacecraft would orbit to make observations of those environments. The project APL and MPIET proposed for the NASA competition was "a very prudent balance of high risk/high gain aspects and somewhat lower gain, but much surer aspects." Krimiges and his MPIET counterpart Gerhard Haerendel found room for five experiments on the American spacecraft and six on the German one.

NASA headquarters assigned responsibility for the project to its Goddard Space Flight Center, which appointed a project manager (an engineer) to oversee construction and integration of the two spacecraft and to disburse funds to the experiments. However, the project manager was not able to exercise the kind of authority held by the Keck project manager, because the science teams were numerous, unified, and (in the case of Krimiges and Haerendel) worked within their organizations with engineers equally experienced in the intricacies of making a spacecraft work. Instead of conceding authority to the NASA project manager (as the University of California and Caltech astronomers had conceded to the Keck project manager), AMPTE was governed as an oligarchy of the scientific leaders who had the support of the participating scientists in the event of any conflict with the NASA project manager. In two important instances, conflict between scientists and project management was resolved in favor of the scientists. The first occasion came when MPIET personnel recognized that the inert mechanical adaptor between the two spacecraft could carry instruments and suggested that the British Rutherford-Appleton Laboratory, with which MPIET had previously collaborated, join the collaboration to build what would, in effect, be a third spacecraft. The project manager objected to expanding scientific objectives and capabilities (and any attendant technical risks) beyond what NASA had approved. "So," the MPIET scientist recalled, "the two of us got into a severe fight about that. Of course, [the lead scientist] was entirely on my side. All the scientists were on my side. But it was essentially a fight between [the manager] and myself, which of course, I won. Because I had the better arguments and we had the mass. But this created tensions which were never quite buried, I would say, through the mission."

The second occasion occurred when Tom Krimiges pushed for the creation of an APL data-management center that could process and combine multiple data streams in near real time. Such a center was not part of NASA's tradition. As one scientist saw it, "it has been a traditional sore point of NASA missions that people clutch their data to their breasts." The center promised to be both risky and costly without improving the capabilities of individual experiment builders to process and analyze their respective data streams. On the other hand, it would enhance the active experiment, which was Krimiges' passion, by enabling scientists to make decisions about the ion releases in response to on-the-spot interpretations of conditions inside and outside the magnetosphere. Again, the manager resisted, at first with the support of some of the scientists: "There was nobody on board with a track record of great reticence in sharing data, and people bought into the concept . . . with some reluctance. And I was originally reluctant, and I think that the fact that I turned around and became a partisan of it also helped convince [the project manager]. . . . We all bought into this concept of a unified science data center, and shared data." Though not consciously entering into back-scratching politics, the scientists who joined the collaboration later sensed that they should make common cause with the instigators. So all supported one person's desire to maximize the value of ion releases, another's to test a model for the origin of charged magnetospheric particles, and another's to collect measurements in the natural environment. Scientists could keep the project manager from influencing the scientific capabilities of the project while still benefiting from his well-recognized ability to spot and correct flaws in the details of spacecraft construction.

Thus, through growth, AMPTE developed an oligarchic decision-making structure of three scientific leaders over matters that were as much technical as scientific. The project manager and his corporate superiors at NASA headquarters did not exercise nearly as much authority as Keck's project manager and board of trustees.

Conclusion

There is undeniable appeal to the idea that multi-organizational collaborations should be "just large enough" in size and duration to accomplish well-defined objectives with budgets and participants known in advance. This would provide a basis for evaluating the economy and performance of collaborations under the premise that smaller is better. But appealing ideas are not always good ideas. Smaller is better in the obvious sense that it is better to use no more people or time than needed to reach some goal. But smaller is not better in the sense of better conceived, better formed, better managed, more satisfying to participants, or more serviceable to the relevant scientific communities. Objectives change, along with budgets and participants. Most collaborations could be smaller or larger without engendering major structural change.

Minimizing magnitude need not and should not be an end in itself. There are benefits to increasing the size of a collaboration *and* waiting for funds. Expanding a small collaboration is not always *just* a matter of currying political favor with a broader range of scientific communities. AMPTE, which was initially conceived as a small collaboration to perform an active experiment and little else, became scientifically more capable and sophisticated as well as politically more feasible because its instigators kept adding participants while waiting for an opportunity to submit a proposal to NASA and then for NASA to release funds. Politically expedient expansion can itself be an impetus to increase scientific opportunities and to enliven a scientific community. IRIS did not need the seismologists of its 22 founding organizations—let alone the 90 more it eventually acquired—to supervise three technical teams that contracted out for the instrumentation and software needed for a global network of digital scismometers and a large stable of digital portable seismometers. But the political necessity of broad expansion in organizational participants ensured that the collaboration's facilities would be developed to meet the needs of a broad swathe of academic geophysicists.

More individual participants, more organizational participants, longer lives, and bigger budgets do push collaborations toward elaborate bureaucracy. As more participants get involved, oligarchies emerge to make decisions about resources. As a project's duration increases, boards of non-participants arise to review collaboration work. Fortunately, these features do not generally make collaborations inhospitable to the kinds of social relations and professional practices preferred by scientists. Deliberations over the interpretation of data may remain participatory while an oligarchic subset of the collaboration considers the budget. A research team or member organization may be able to plan its uses for a collaboratively acquired facility while an external panel reviews the facility's engineering. The next two chapters elaborate on how multi-organizational collaborations create scientific opportunities without violating their participants' sense of traditional values for science by making collaborations as non-collaborative as is possible.

3
Organization

In his history of the UA1 and UA2 particle-physics experiments at CERN, John Krige exploits the "tenacious image" that besets accounts of collaborative projects in physics. That image is a contrast between individualized and bureaucratized science. On the one hand, there is the scientist as an autonomous craftsperson who controls all the tools needed to create new knowledge, with free rein to use those tools in experimental demonstrations for other scientists. On the other, there is the scientist as a factory worker—a part of a multi-layer managerial structure that emulates an industrialized workplace—consigned to contribute only a specialized segment to a larger project and unable to work independently on experiments of one's own design. Hierarchical relationships replace the "free exchanges among equals," "bureaucracy is rampant," and "decision-making processes have become increasingly formalized" (Krige 1993: 234). At the extreme, such activities are boring, exploitative work that provide little scope for creativity and alienate scientists from research and the new knowledge it produces. The "free-wheeling, creative atmosphere of the university laboratory has been supplanted by the constricting procedures and regimentation of the large corporation" (ibid.: 254).

Krige finds the contrast inappropriate and unenlightening for understanding particle-physics experiments. Our view, based on a broader survey of collaborative sciences, is that the image is useful only when properly qualified by examining the amalgam of features that are associated with "bureaucracy."[1] Max Weber's classic characterization of bureaucracy (1946) specified the presence of a division of labor, hierarchy of authority, written rules and regulations, and a principle of technical expertise.[2] Weber's successors in organizational theory have had a harder time of it, as they came to understand that an "ideal type" is just that. Social formations are not necessarily bound to a specific configuration of features.

The components of bureaucracy must be examined individually with an eye to variation in the extent to which each may or may not be present.

To complicate matters, while bureaucracy was originally conceived as a rational and efficient form of organization, many of its current connotations involve unnecessary formalization, the waste of time and resources, the proliferation or rules and "red tape." These bureaucratic pathologies have been widely recognized since the 1940s. (See, e.g., Merton 1940.) However, this chapter deals with "normal" bureaucracy in scientific collaborations—that is, with the internal organizational and managerial mechanisms of multi-institutional research enterprises. Regrettably, although there is a vast literature on inter-organization relations, organizational studies have largely ignored scientific collaborations as objects of inquiry,[3] focusing on production (Pfeffer and Salancik 1978; Browning et al. 1995; Gulati 1995; Powell et al. 1996), service (Alter and Hage 1993), government (Clarke 1989), and non-profit organizations (Kang and Cnaan 1995). In appendix A, we review several of these recent approaches.

Of particular interest for social studies of science is Karin Knorr Cetina's ground-breaking work on high-energy physics. Knorr Cetina (1999) characterizes collaborative experiments as post-traditional communitarian formations, with object-centered management, collective consciousness, and decentralized authority. Such "mega-experiments" introduced a new form of collaborative work predicated on collectivism, the erasure of the individual epistemic subject, non-bureaucratic work mechanisms, the lack of formal structures, and absence of fixed internal rules (ibid.). Such a flexible, democratic, and interdependent organizational configuration is the antidote to the hierarchy and control that might otherwise accompany the move toward "Big Science." Our data from high-energy physics experiments was collected on experiments that occurred before the 1990s. But the organizational arrangements are broadly similar. Were these arrangements the harbinger of the future, as collaborations in other fields became larger as well, or perhaps by serving as a model for the design of other collaborations? Our results suggest the contrary, that particle-physics collaborations are exceptional.

Particle physicists have employed a distinctive form of organization. American particle physicists enjoy a relatively uniform infrastructure of funding agencies and accelerator laboratories.[4] During the period covered by the present study (for particle physics, about 1975–1990), electronic detectors were built at accelerator laboratories to conduct experiments.[5] Competition for time and space at accelerator laboratories, routinized institutional politics, and the limited range of experimental styles heightened the competition for discoveries and theory testing. These conditions imposed extraordinary discipline that pushed collaborators to adopt similar organizational structures, granting broad rights of participation to all members of the collaboration, from graduate students to senior faculty. Such Athenian-style democracy has produced remarkably successful outcomes. Yet when we set aside preconceived notions of disciplinary peculiarities and investigate a broader sample of physics collaborations, a narrow focus on particle physics as a model for collaboration turns out to be misleading. This is only one of several possible organizational formats. Equally important, it is the only field-specific arrangement. A variety of more overt formal structures characterize collaborations in other areas of physics and in other sciences as well.

The ideal types of autonomous craftsman and factory worker are unlikely to be encountered by historians, ethnographers, managers, and policy analysts. They are extremes in a spectrum whose mid-ranges are important to delineate and characterize because it is where most collaborations lie. We do so by differentiating collaborations along four dimensions: (1) the extent to which they employ formal rules and documents, (2) their use of a specialized division of labor to carry out research, (3) their decision-making hierarchy, and (4) the degree to which a scientific leader sets research directions. In the next section, we elaborate these dimensions through evidence from our qualitative interviews. Then, as in chapter 1, we use cluster analysis to generate a typology of the ways collaborations organize and manage themselves. This analysis shows that multi-organizational collaborations display patterned diversity, mixing and matching the features of classical bureaucracy, and indicating that there are many ways of organizing. In our sample of collaborations, four main types can be identified. As always, taxonomy is a compromise between an elegant but simplistic appeal to diametrically opposed ideal types and an empirically unassailable but conceptually useless focus on the traits of individual collaborations.

With the exception of particle-physics collaborations, which dominate a single category, there is no clear-cut relationship between the way collaborations are organized and their disciplinary specialty—even particle-physics collaborations can be found in all categories. Further, collaborations from other specialties may be found in the category dominated by particle physics. In space science, the NASA and ESA space-flight centers have always managed collaborations, but the powers of flight-center-appointed project managers and project scientists have varied significantly. In geophysics, no single American funding agency dominates the specialty (as NASA does in space science). No institutions comparable to space-flight centers exist to manage geophysics collaborations. The organization and management of geophysics collaborations fell into patterns according to whether the collaborations "imported" industrial techniques into academic geophysics or "aggregated" measuring techniques established within academic geophysics. Some geophysics and space-science collaborations more strongly resemble each other than collaborations in their respective disciplines. In short, *disciplinary traditions and culture are not of overriding importance* to the organization and management of multi-organizational collaborations. Within every discipline studied, the organizational and managerial needs of collaborations spanned a broad range. Yet the ranges for each discipline are similar, reducible to four distinct types of collaborations: bureaucratic, leaderless, non-specialized, and participatory.

Dimensions of Collaborative Organization

The major dimensions of organizational variability in collaborations include division of labor, system of scientific leadership, degree of formalization (including written contracts and administrative leaders), and decision-making hierarchy. Do the teams in a collaboration work on similar tasks, or on different tasks? What configuration of leadership is involved? Do participants distinguish sharply between intellectual and administrative matters? Are managerial arrangements stipulated in formal contracts and documents? Are specific rules and procedures established for reporting and reviewing tasks? Is there flexibility in the timetable? Are decisions made by the collaboration as a whole? Is a committee designated, inside or outside the collaboration, to evaluate its work?

The division of tasks in collaboration may be more or less specialized. The sociologist Emile Durkheim distinguished between societies on the basis of whether their division of labor produced mechanical solidarity or organic solidarity (1933 [1893]). The former was said to characterize pre-modern or primitive societies (in which most people tended to engage in the same basic kinds of activities), the latter to characterize modern societies (in which work was highly specialized and members were interdependent). The distinction works even better for scientific collaborations of several organizations. In most projects, each team has differentiated tasks or functions. The leadership seeks to integrate interests and competencies, manage their relationships, and keep the project headed in the right direction with minimal delay and conflict. In a minority of projects, teams had similar tasks or functions and the purpose of the collaboration was to aggregate team efforts. For example, a collaboration that conducts clinical trials of medical instrumentation may require all participants to use the same diagnostic protocol so their data may be aggregated into a pan-collaboration database.

Usually an organization has a single position at the apex of its hierarchy. But in about 20 percent of the collaborations in our sample there was no single scientific leader—that is, no scientist viewed by others as inspiring the collaboration intellectually, actively managing resources, and making judgments on behalf of the collaboration.[6] In 70 percent of the cases, there was a distinct administrative leader—an engineer[7] who managed the collaboration's resources, or who oversaw the assembly and integration of its instrumentation.[8] About half of the collaborations in our sample exhibited both types of authority, scientific and administrative.[9]

Plasticity in the leadership structure of collaborations originates in struggles over just how unified collaborators wish to be, given the voluntary and temporary character of collaborations. As members thrash out the jurisdictional boundaries among the collaboration as a whole, project teams, member organizations, and individual participants, they are often uninhibited by the norms of organizations in other domains. Instead, they argue for designing their preferences into collaboration—knowing that other members will want to meet enough of their demands to keep them from walking out. They can afford to make concessions because they will not have to live within the collaboration forever. The result is diversity in leadership structure as well as in the kind of person who is deemed qualified to exercise leadership.[10]

The design of a leadership structure is always one of the most significant issues in the life of a project.[11] A collaboration with scientific but not administrative leadership arose from scientists who possessed a vision of shared instrumentation but nevertheless wanted their separate organizations to control the collaboration's resources:

We never did get comfortable with the idea of having a real project manager for several reasons. For one thing that person would then have to be given control over resources of the other institutions, and people were never really comfortable with that. And another reason is that then if you're going to hire a project manager, such people aren't cheap, and that would just take away from resources which might go directly to the [research and development]. So there was always a conflict about do you get enough benefit from going out and hiring a real project manager as opposed to sort of bootlegging it off the top of other duties.

The absence of a top manager did not, by all accounts, indicate an absence of structure. In fact, there was a board of directors (with representatives from each institution, proportionate to their resource investments), and an executive committee (including the scientific leaders of all the organizations) served as the "bootleg project manager." This collaboration had multiple scientific leaders but no administrative leader.

In contrast, each space-science collaboration had a project manager to oversee the design and construction of the spacecraft, including the payload of scientific instruments. But such collaborations may have no scientific leader if participating scientists wish to maintain their autonomy and deal with their conflicts case by case. In these cases, the project scientist, though seemingly parallel to the project manager, does not drive the collaboration intellectually or control resources:

> You have some very powerful principal investigators [in this collaboration].... On the other hand, in [the space-flight center] you have a young, relatively inexperienced scientist [as project scientist], and then you cannot expect the young, relatively inexperienced scientist to be more dominant than [these PIs]. The possibility that the project scientist has is through persuasion and good argumentation, [especially in] cases where two PIs are in conflict with each other.

Like other space science projects, this collaboration had only an administrative leader and questions of rank and boundary between administrative and scientific authority were moot.[12]

About half of the collaborations in our sample had both kinds of leaders. In such cases we routinely asked about their relationship. What were their relative positions and how interdependent were they? In many of these collaborations, one leader could impose his will on the other. In the large majority of these, administrative and scientific leaders had clearly delineated duties.[13] One lead scientist described the duties of the engineering leader when the latter owed his position to the former[14]:

> What he has to do is basically look over the whole project time line and let me know when he thinks somebody is falling off. He has to talk to everybody to make sure everybody fits together ... and that no one is violating any safety issues, and that basically the experiment will be coherent mechanically. So far, he's done a fine job in basically isolating me from nitty-gritty kinds of things. No one is complaining to me about him, either, so I assume he's doing a good job.

This lead scientist felt authorized to hold the engineering leader accountable, but he was not sophisticated enough about mechanical engineering and safety regulations (which were in the engineer's domain) to fill both roles himself. Occasionally the engineering leader had authority over the scientific leader. The scientists in the collaborations usually resented that:

[When we scientists] were just about ready to try [a novel technique], the project [manager] would run into trouble in some other aspects, so they just wiped that [part of the budget] out. Something that could have been used to improve science was just used for other needs.

The lament of the lead scientist was not that the project manager was overstepping his expertise, but that the project manager had the power to allocate collaboration resources and thus check the demands—here, the *changing* demands—of scientists in the several institutions that had a larger role in the initial design of the project. The existence of a strong leader is a double-edged sword. Collaborations with scientific and administrative leaders of equal rank resulted when projects that already had a leader recruited an independent person in the other role. Such situations occurred when conflicts among scientists created problems that an administrative leader could not solve, or when technical difficulties in collecting data behooved a scientific leader to share power with someone better able to deal with the technical problems. An example of the former occurred when a NASA flight center convinced a prestigious university scientist to serve as project scientist rather than one of its in-house scientists. In the eyes of the project manager, a powerful project scientist made managing easier:

Prior to X all the project scientists that we had [flight-center] full-time people. And while they did okay, they . . . certainly weren't bona fide leaders of the science working group, because they did not have the stature or the credentials with the principal investigators. That changed with [this project]. [This project] brought X, who was a professor . . . , who had built science instruments, and who was highly regarded. . . . That was a masterful stroke.

Technically, this project manager may have had the authority to make a final decision in a dispute over intra-collaboration resources. But he was also clearly happy to defer to the project scientist, and thus not run afoul of any of the principal investigators, especially when the dispute only involved PIs. The two positions were essentially equal in authority.

An example of scientists creating a position of administrative leadership occurred when a university scientist, who had assembled a collaboration of university and corporate scientists to use a national facility, realized that only an on-site scientist could cope with the difficulties of making the facility perform as the scientists wanted:

Brookhaven encourages outside users, so X [a Brookhaven scientist] was happy when somehow a consortium of [outside] people . . . wrote a proposal to NSF. It was Y's [the scientific leader's] grant, but it was basically to pay for X to develop the positron beam on the reactor floor, operations level of the reactor, at Brookhaven.

The scientific leader was the principal investigator on the grant from the outsiders to NSF. The Brookhaven scientist was the administrative leader, because he was responsible for assembling the instrumentation at the research reactor. Their authority was equal because they were both independent scientific users of the instrumentation.

The exercise of leadership within an organizational structure is not the only form of control employed by formal organizations. We inquired into the use of other procedures associated with bureaucratic organizations: formal contracts that specify roles and assignments, well-understood rules for reporting developments within the collaboration, rules for reporting developments outside the collaboration, and hierarchical procedures for making decisions on collaboration activities. But as with leadership, multi-organizational collaborations use arrangements that would be untenable for permanent organizations. Consider, for example, the use of formal contracts. Even if "contract" is defined so loosely as to mean any written document that stipulates the obligations and privileges of members, only two-thirds of the collaborations utilized formal contracts. Even these contracts were not always significant features of these collaborations:

There was only a contract drawn up between the [facility] and our research team, but that contract actually was supposed to have been drawn, of course, at the beginning of the collaboration. But it was not drawn up until basically the whole project was completed. So what that tells you is the very informal nature of this project.... People honor each other's promises and it's basically done verbally.

Notably, this speaker is in the private sector, ordinarily thought to depend *more* on formal contractual arrangements than universities do.

Most collaborations had a system of rules for reporting on intra-collaboration work. In the absence of powerful, unified leadership, such rules—in combination with individual competitive pride—can be the principal source of accountability. No one wants to be the bottleneck in the eyes of fellow collaborators, the obstruction to the accomplishment of project goals. Such a strategy is both common and successful in particle-physics collaborations, which labor under the external discipline of both the accelerator schedule and competing collaborations, but could break down in collaborations that were developing autonomous or unique research facilities.

The final aspect of organizational structure is decision making, ranging from hierarchical to consensual. What seemed clear in many interviews is that informants seek to emphasize the consensual aspects of decision-making practices that might, at best, be characterized as a blend of consensual and hierarchical. This is often reflected in the degree to which leadership subgroups dominated decisions concerning scientific, engineering, or administrative matters, such that consensual decision making operates at the top level while a more hierarchical structure characterizes the organizations themselves[15]:

> When we have to make a decision that involves the entire collaboration, we do it mostly amongst the PIs. So there is no one person who makes the final decision. It's a consensus among the PIs, and then if we need advice, we'll consult the people in the consortium, the students and the postdocs. So I guess at some level you'd call it a two-tiered hierarchy.

Types of Collaboration

To generate a taxonomy for the bureaucratic character of collaborations, we performed cluster analysis on the organizational dimensions of collaboration—formalization, hierarchy, scientific leadership, and division of labor.[16] While the dimensions themselves are based on theory, they are also shaped by the empirical associations evident in the data.[17] Any technique designed to assess the similarity or proximity among a set of objects has the potential to classify those objects differently depending on the dimensions used for the classification. The general aversion among qualitative investigators to categorization, to "putting things into boxes," is warranted when a typology is extended far beyond its bounds. In chapter 1 we developed an analytical typology for understanding the formation of collaborations, but we do not expect it to be equally useful for understanding their organization. It is not putting objects into boxes that is objectionable, but leaving them too long.

As we emphasized in chapter 1, cluster analysis only arrays members of our set of collaborations according to their degree of structural similarity, and the classification scheme we impose balances the intellectual economy of a small number of groups against the virtue of creating internally homogeneous groups. Nevertheless, inspection of the dendrogram in figure 3.1 reveals four main groups of projects.[18] The standard deviations for each group are overwhelmingly smaller than the total standard deviation. This indicates that the clusters are quite homogeneous internally and heterogeneous externally, a hallmark of a good classification solution. With one notable exception, organizational types are not field specific, but rather cut across fields. The exception is type 4, which consists mostly of particle-physics collaborations.

96 Chapter 3

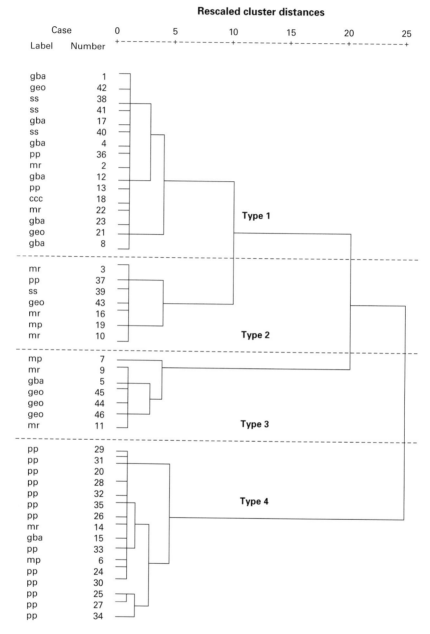

Figure 3.1
Dendrogram showing organizational clusters. (Type 1: Bureaucratic. Type 2: Leaderless. Type 3: Non-specialized. Type 4: Participatory. pp: particle physics. ss: space science. geo: geophysics. mr: materials research. gba: ground-based astronomy. mp: medical physics. ccc: computer-centered collaborations.)

Table 3.1 facilitates the interpretation of the four organization types of collaboration.

We deem the top group "bureaucratic" because it possesses many of the classical Weberian features: a hierarchy of authority, reliance on written rules and regulations, formalized responsibilities, and a specialized division of labor. These features do not necessarily make collaborations rigid; usually their purpose is to protect organizational interests, since collaboration is, by definition, an intendedly ephemeral joining of partners. The next two groups are "semi-bureaucratic," characterized by the lack of some Weberian characteristics and the presence of others. A "leaderless" collaboration also deploys formally organized, highly differentiated structures, but it does not designate a single scientific leader; their administrators attend to external relations and ignore internal politics. A "non-specialized" collaboration possesses lower levels of formalization and differentiation but features a scientific leader who exercises intellectual leadership. A "participatory" collaboration is characterized by the general absence of these Weberian features. In a participatory collaboration, decision making is largely consensual and organizational structure is developed and maintained without recourse to formal contracts or hierarchical management. This is the only type that correlates strongly with a research specialty, and the reason we speak of "particle physics exceptionalism."

Bureaucratic Collaboration

Bureaucratic collaboration is characterized by a high incidence of the classical Weberian features: hierarchy of authority, written rules and regulations, formalized responsibilities, and a specialized division of labor (Weber 1978). Although there are variations among the bureaucratic collaborations we studied, several manifestations of this organizational pattern are common: extensive external evaluation, committees upon committees with various designations and functions, officially appointed project managers, clear lines of authority (administrative and scientific), and a well-defined hierarchy. Such a set of characteristics originates with the need to make sure that no single organization's interests dominate the collaboration, but it also befits multi-organizational collaborations whose scientists can sharply distinguish the collaboration's "engineering" from its "science." In these cases collaborators can pursue science autonomously from the engineers (and often each other)—provided the engineering is well done and competently documented.

Table 3.1

Average characteristics for each cluster: formalization, hierarchy, scientific leadership, division of labor. The "formalization" index has a minimum value of 0 and a maximum value of 1.20. The "hierarchy" index ranges from 0.75 to 2.50. "Scientific leadership" ranges from 0 to 2 (more than one leader). Division of labor was coded 0 for unspecialized and 1 for specialized.

	Formalization		Hierarchy		Scientific leadership		Division of labor	
	Mean	S.D.	Mean	S.D.	Mean	S.D.	Mean	S.D.
Type 1: Bureaucratic ($n=16$)	0.88	0.16	2.02	0.41	1.25	0.45	1.00	0.00
Type 2: Leaderless ($n=7$)	0.74	0.15	1.64	0.66	0.00	0.00	0.93	0.19
Type 3: Non-specialized ($n=7$)	0.66	0.37	1.66	0.40	1.14	0.69	0.07	0.19
Type 4: Participatory ($n=16$)	0.21	0.20	1.10	0.37	1.12	0.62	1.00	0.00
Total ($N=46$)	0.59	0.36	1.59	0.58	1.00	0.67	0.85	0.35

This type of arrangement is illustrated by the collaboration between the University of California system and the California Institute of Technology (with secondary participation by the University of Hawaii and NASA) to build the Keck Observatory. In the late 1970s, University of California astronomers responded to encroaching light pollution at UC's Lick Observatory by entertaining novel telescope and observatory concepts that would vault them into the vanguard of optical observing power. They calculated that a telescope with the largest primary mirror ever constructed, if placed at an optimal site and accompanied by state-of-the-art detecting instruments, could yield as much as 100 times the observing power they currently enjoyed. The chancellor of the University of California provided seed money to enable proponents to flesh out their ideas and organized a formal review of the resulting proposals. The result was an endorsement for the idea, championed by Jerry Nelson of the Lawrence Berkeley Laboratory, of an optical telescope with a segmented 10-meter mirror that would be unprecedented in size and would require a novel operating mechanism.[19]

The technical challenge of the mirror plus the overall size and sophistication of the observatory promised to make the project extraordinarily expensive. Quite simply, UC needed partners. With the support of UC astronomers, UC administrators contacted the California Institute of Technology, UC's major competitor in astronomy. Though UC intended Caltech to be a junior partner, Caltech ended up as an equal because, while UC's fund-raising luck turned sour, Caltech found William Keck. Keck provided the money needed for a complete observatory, insisting that it be named for him. The University of Hawaii became a junior partner on the strength of contributing a site with observing conditions that put the mirror in "best light." The Keck Foundation subsequently offered to provide two-thirds the cost of a second telescope that would make optical interferometry possible. NASA became a junior partner by providing the final third. The total funds invested in the construction of the twin telescopes exceeded $140 million.

The quantities of money involved and the history of competition between Caltech and the UC campuses induced the university administrators to formalize their arrangements explicitly in order to guarantee that no single university imposed its interests on the collaboration. They created a corporation, the California Association for Research in Astronomy, whose sole purpose was to design, build, and operate the Keck Observatory. The documents of incorporation specified fiscal responsibilities and payment schedule, isolating CARA from the budgetary politics of any member organization. The documents set forth a hierarchical authority structure for the collaboration and specified the general responsibilities of each part of the structure so as to ensure that the major member organizations had equal authority.

A board of directors, with equal numbers of representatives of UC and Caltech, was the ultimate authority and actively supervised the work. It met every 2 months to check on progress, commissioned external panels of experts to review developments at strategic points, and required its approval for any significant budgetary shifts within the collaboration. The staff of CARA was responsible for the things any optical astronomer, regardless of further scientific or technical specialization, needed from the observatory: the telescope (including the mirror and the pointing and tracking mechanics), observatory buildings, the infrastructure to operate the telescope, and a suite of scientific instruments. CARA's administrative leader was the Project Manager, Jerry Smith, an engineer from the Jet Propulsion Laboratory. Smith was the unambiguous decision maker on all issues faced by his staff: "CARA is absolutely hierarchical. There's a manager. He makes the decisions. There's no negotiation about it. There's no 'what do we all want to do.' There isn't any 'we.' There's only the director or the manager." Finally, a Science Steering Committee (SSC), made up of astronomers from the universities, was responsible for producing a set of scientific instruments and for advising the board of trustees and the CARA staff about engineering options that could affect the Observatory's scientific capabilities. The SSC made its internal deliberations consensual on the assumption that it would not encounter issues that required an exercise of authority to resolve. But the SSC's chairperson (who changed over the course of the project) and the Project Scientist (the chief author of the segmented-mirror concept) were the authoritative conduits of messages from astronomers to CARA and the board. These arrangements successfully eliminated the interests of the individual organizations from observatory policies and eliminated administrative ambiguity from the collaboration. Individual participants disagreed strenuously over the substance of observatory design and construction, and their arguments may well have been enflamed by personality conflicts, but there was no doubt about the proper procedures for reaching decisions or the good faith of the decision makers:

[Anything] having to do with money formally went to the board of directors for approval. It was clear and unambiguous what the lines of authority were on really every issue, and I think it worked fine. And the organization generally speaking did not pay much attention to UC versus Caltech. It was a contract, but where the money came from, but that was a contract, that was done, there was nothing to talk about once the contract was signed.

What is important about this quotation is its source, a partisan in the conflicts, who believed the wrong side won more often than not. But they won "fair and square," through legitimate decision-making procedures under a hierarchical authority structure. Through clear and formal specification of the division of responsibilities for each structural component, traditional academic rivalries were managed as the project moved toward completion. This project did not so much balance the interests of its independent organizations but rather banished the interests of the constituent organizations by creating a formal Weberian bureaucracy.

Leaderless Collaboration

Leaderless collaboration, like bureaucratic, has formally organized, highly differentiated structures that serve much the same purpose: to ensure that private interests are not stamped on the collaboration, especially when high levels of resources are at stake,[20] and to ensure that appropriate people stay focused on specialized tasks. But unlike bureaucratic collaboration, leaderless collaboration does not designate a single scientist to decide scientific issues or even to represent scientific interests. The strong sense of hierarchy present in Keck—in which some scientists were more important than others, the important scientists felt they were outranked by project management, and the board of directors actively monitored developments and adjudicated disputes—was not present in leaderless collaborations. In leaderless forms, administrators sought the input of research scientists on important matters, appointed scientists to take charge of developing instrumentation, and attended to the collaboration's external relations while benignly neglecting internal politics.

The DuPont-Northwestern-Dow Collaborative Access Team (DND-CAT) is a good illustration of this style of organization.[21] Initially, DuPont and Northwestern University (the Dow Chemical Company joined later) agreed to build a beamline at the Advanced Photon Source at Argonne National Laboratory for various kinds of materials, chemical, physical, engineering, and biological research. The two organizations differed in their ability to capitalize the collaboration and in their needs for proprietary and published results. Lacking any history of collaborating on this level, they spelled out their rights and responsibilities in a legally binding agreement that was just short of formal incorporation[22]:

[It was] the kind of thing one would create for a joint venture with two companies, where you set up a, it's almost like a separate little incorporated outfit. It's not incorporated. It's very close to being a free-standing organization with every aspect—intellectual property rights, all the financial parts, the duties and responsibilities of the board, the director, all of the details of how you get out of this agreement, what you owe.... It's a very thorough document.

As with the Keck Observatory collaboration, the legal agreement insulated the collaboration from members' budgetary politics by stipulating the schedule and amount of payments each member organizations would make to fund the collaboration. As with Keck, ultimate authority was vested in a board of representatives of the member organizations, and the board was to ensure the collaboration did not become an extension of the interests of any single member.[23] As Keck's member organizations entrusted the design, construction, and operation of the telescope and the observatory's infrastructure to CARA, so DND-CAT entrusted the design, construction, and operation of the bulk of the beamline to a professional group with the *de facto* responsibility of serving the collaboration.[24] And just as Keck commissioned scientists at its member organizations to design and build instruments for use in conjunction with the telescope, so DND-CAT relied on scientists at member organizations to design and construct "end station" instrumentation to be used in conjunction with the beamline.

The formal contract defined a well-understood system of responsibilities and reporting. The DND board, like the Keck collaboration, controlled the budget. However, the relationships among board, staff, and scientists at member organizations were quite different. The Keck scientists were all astronomers with ideas for the design of an observatory and the interface between a telescope and instruments placed at the focal point. In contrast, DND was to serve a multi-disciplinary set of scientists who, in many cases, had not previously used synchrotron radiation in their experiments. Keck needed to impose an authority structure on its scientists to ensure they would not make conflicting requests of CARA staff, and the Keck board had to involve itself in technicalities to decide when it was feasible for CARA staff to take scientists' requests and suggestions in addition to their burden of building an unprecedented telescope. The DND staff, in contrast, did not need protection from the board because the beamline was not novel in design (though some of its components stretched the state of the art) and its users were as much in need of learning about synchrotron radiation from the staff as they were in need of making their expectations known to the staff. Instead of a single Science Steering Committee to channel communication to DND staff, DND had several science working groups that each interacted with the staff to learn about the beamline's capabilities and how they could be used. So long as collegiality prevailed between staff and working groups, and so long as the staff could meet the technical burdens they assumed within the limits imposed by the collaboration's budget, the board was passive. Indeed, board meetings became so mundane that it began to operate by teleconferencing. DND operated by formally establishing differentiated roles and leaving the specialists to work and interact without a hierarchy of leaders to channel communication and adjudicate disputes.

Non-Specialized Collaboration
Non-specialized collaboration is the complement of leaderless collaboration. Whereas leaderless collaboration is bureaucratic in formalization and differentiation, non-specialized collaboration is bureaucratic in hierarchical management and leadership structure. Non-specialized collaboration exhibits less formalization and differentiation in member tasks than either bureaucratic or leaderless collaboration.

An instance of this pattern is the International Satellite Cloud Climatology Project (ISCCP), which used radiance data from weather satellites to generate statistics on the global distribution and characteristics of clouds. Cloud-radiation feedback, along with the ocean's circulation, had long been viewed as a major source of empirical uncertainty confronting climate modelers. However, until the International Council of Scientific Unions (ICSU) and the World Meteorological Organization (WMO) jointly created the World Climate Research Programme (WCRP), scientists with climatological interests were inhibited both by the technical difficulties of collecting and processing global data and by the political difficulties of acquiring support when national governments were most concerned with localized or shorter-term phenomena relevant to weather prediction. In the late 1970s, an international band of atmospheric scientists, with the help of computation experts, began investigating whether model-relevant global cloud statistics could be obtained from the information that weather satellites produced, even though those satellites collected greater quantities of lower-quality information than climatologists would have liked.

The prospect of addressing a major scientific issue without undertaking research and development of dedicated instrumentation had obvious appeal to the fledgling WCRP. It sponsored a string of workshops in which atmospheric and computer scientists produced concrete plans for sampling and calibrating data from geostationary and polar-orbiting weather satellites, applying an algorithm for obtaining information on the presence and characteristics of clouds from the raw data, and finally passing the processed data to a central data archive. In 1982, on the strength of these plans, WCRP formally made ISCCP its first project.[25] The international group of scientists who instigated ISCCP was collectively inclined to disperse responsibilities for the project among themselves and to rely on their shared scientific motivations and standards to guarantee the quality of the work each would perform. During one of the workshops preceding ISCCP's formal start, the scientists drew up a view graph for how the project would work: "What was interesting was that at that stage we didn't imagine a central processing center; we imagined distributing processing. Everybody would get a copy of the code and apply it locally and ship the results to some central location." The scientists did need enough formality in their arrangements to satisfy the requirements of an international agency, whose imprimatur was essential for the project's scientists to convince their national governments to contribute the data from their weather satellites and the funds needed to process the data. But they cared so little about the formal documents that major participating scientists did not even attend the WMO/ICSU meeting at which the representatives of national governments made their pledges to ISCCP.

However, the scientists quickly came to realize that, while informality was appropriate to their sense of community, decentralization undercut their ability to satisfy their scientific objectives. Their main need was to agree on a single algorithm for deriving characteristics of clouds from the sampled and calibrated data. Before the formal start of ISCCP, they realized that they lacked intellectual consensus over the algorithm: ". . . the last decision we made at Hamburg [the site of the last workshop before the formal start of ISCCP] was that we didn't know how to analyze the data. In other words, we had all the scientists there fighting and yelling and screaming and disagreeing, and the bottom line was that there wasn't any clear choice [among their algorithms]." Reaching and implementing a consensus on the algorithm was impossible without centralized authority and resources. Even a realistic, fair comparison of the various algorithms proved elusive because the advocates of various approaches to an ISCCP algorithm were not all equally capable of handling data sets that were large enough in size to be a fair test of what an ISCCP algorithm faced. If scientists with pride of authorship in their algorithms varied in their ability to subject them to competitive tests by processing the same data, *a fortiori* how was the project to impose a uniformly high standard on the teams that were going to sample, calibrate, and process the data the project assigned them? A team could not much advance its own interests, the collaboration's interests, or even the interests of the broader scientific community by lavishing specialized craftsmanship on a task that was only valuable insofar as others did it just as well, and in just the same way.

Decentralization was therefore not viable, because the collaboration could not rely on either the narrowly or the broadly defined interests of its members to ensure the teams would perform to their best capabilities. Having already shunned formalization as a source of collaboration-wide discipline, ISCCP scientists were left contemplating the somewhat bitter truth: ". . . in this kind of environment where things aren't really that formal, you don't have much control. So if a center is either not doing the job they said they would do on the schedule they agreed to do it on, you can't do anything because you're not paying the bills." The organizational solution to varying levels of resources, differing approaches to an algorithm for deriving the physical parameters of clouds from weather-satellite data, and the difficulties of working with data from instruments that were not designed with the project's purposes in mind was to centralize operations in a member organization that could then be responsible for adhering to standards: "The data is such a mess—with the instruments misbehaving and calibrations fluctuating and everything else—that it really required a central place where you look at all of it together to try to figure out what the hell it's doing." The question of which organization was to be central was decided pragmatically. Among the agencies interested in supporting ISCCP, only NASA was prepared to support a global processing center. Among American organizations showing interest in processing some of the data, only the Goddard Institute for Space Studies (GISS) in New York was prepared to take on the task.

ISCCP remained without an administrative leader. Managerial duties were performed by scientists from the nations and agencies whose weather satellites were tapped for ISCCP's data. However, GISS's role as the global processing center did make the GISS scientist most involved in ISCCP, William Rossow, the *de facto* scientific leader. Because it made no sense to analyze the data before checking its quality and no sense to distribute the corrected raw data sets for analysis only to recollect them in a central archive, Rossow, by virtue of his willingness to take on the problems of guaranteeing the quality of ISCCP data, acquired authority over the development of the algorithm that made the weather data climatologically relevant. Scientists at other participating organizations each prepared data from the satellite(s) their organizations tracked and shipped the data to GISS for processing through the ISCCP algorithm.

ISCCP and the other collaborations in this category were distinctive in their combination of a lack of a specialized division of labor and a hierarchical authority structure. Each ISCCP team performed the same function—preparing data from a weather satellite for shipment to GISS—and waited for the return of a global data set of cloud characteristics that all could use for climatology. ISCCP internalized professional rivalries among atmospheric scientists, and those who disagreed with GISS decisions, like Keck scientists who disagreed with CARA decisions, faced an uphill battle against a hierarchy. In principle, ISCCP dissenters could appeal to the WCRP program manager and advisory committee, but the burden of proof would plainly be on those suggesting that ISCCP would be better off if GISS changed course.[26] In practice, ISCCP dissenters have mostly honed their approaches to inferring cloud characteristics and waited for the next opportunity to collect data.

Participatory Collaboration

Participatory collaboration is characterized by the absence of the features associated with classical Weberian bureaucracy. This type is the only one whose membership is dominated by a single specialty.[27] Among all the specialties in physical research we examined, particle physics alone has a distinctive "style" of collaboration that makes it exceptional in style of governance and breadth of operations. Participants generally describe their projects as highly egalitarian in governance. Compared to collaborators in other disciplines, particle physicists view decision making as participatory and consensual, define organizational structure through verbally shared understandings or non-binding memoranda rather than formal contracts, and create flat structures with few levels of internal authority. At the same time, the scope of particle-physics collaborations encompasses nearly all the activities needed to produce scientific knowledge, including those activities most important for building a scientific career. These collaborations collectivize the data streams from the individual detector components built by the participating organizations. They frequently track who within the collaboration is addressing particular topics with the data. They routinely regulate the external communication of results to the scientific community.

These features are illustrated by Fermilab 715, previously described in chapter 1, one in a string of fixed-target experiments based on a core membership's ability to produce a beam of hyperons (FNAL 497, FNAL

715, and FNAL 761). Several of the major players had worked together previously, ensuring continuity. Pre-existing relationships were strong and communitarianism was the order of the day (Knorr Cetina 1999). The collaboration obtained data and published results with minimal formalization. No formal contracts were signed despite the fact that this was an international collaboration involving physicists from two Cold War adversaries. The only document drafted was the standard memorandum of understanding between the participants and the Fermi National Accelerator Laboratory—not in any sense a legal agreement.

Collaborating organizations did not pool funds, with the implication that they did not need formal rules to ensure that members received their fair share. Rather, each major American organization had its own contract with DOE or NSF, while the Soviet government supported the participation of the Leningrad group. The collaboration did depend on the ability of Soviet participants to travel to the United States, so by not pursuing a formal agreement, the collaboration effectively gambled that neither government would prove an obstacle. At times it appeared the collaboration might lose the gamble, but each time participants were able to secure the cooperation of the state apparatus.

No administrative or engineering leader was needed in the context of a well-understood division of labor and the absence of sophisticated systems-engineering issues. Most participants were recapitulating or building on past successes that were the foundations of their scientific reputations. Interviewees effortlessly rattled off who was responsible for what:

> The TRD [transition radiation detector] was clearly Leningrad's, since it was their invention. The drift chamber system . . . built for 497 [an earlier experiment], and the design of the channel was the same as we used for 497. So X was naturally responsible. The high resolution proportional chambers had been a Yale responsibility for 69, so it was natural for Y to take that over.

The only desired component for which there was no experienced team was a lead glass array, which was built by Roland Winston's team at the University of Chicago. In view of the university's reputation and Winston's level of interest in the experiment, that was hardly cause for concern.

The experiment did have a designated scientific leader, or spokesperson, but no hierarchy of participating scientists. Once Fermilab approved the proposal for beamtime, the spokesperson's duties were a mixture of scientific coordination (which was intellectually strenuous, since it required he know the technical needs and characteristics of each component) and administrative routines (ensuring that the collaboration covered all data-taking shifts). However, when the collaboration met as a whole to discuss the operation of the detector, the combination of data streams, or the reliability or significance of findings, all titles disappeared. Not even the most vituperative of Cold War rhetoric put a damper on the unrestrained, egalitarian discussions of the experiment:

> It was entertaining to watch. . . . The Russians first came shortly after Reagan's speech in which he declared the Soviet Union the evil empire. They were understandably circumspect and a bit clannish in general. . . . We'd finally sit down around the table and start to discuss physics and that evaporated. On a given day, Chicago and Yale would gang up on Leningrad and Fermilab and on the next issue they would change sides, they would split. It was the usual physics free for all, as in all collaborations.

Thus, even strong cultural and ideological differences did not inhibit these physicists from a participatory exchange of scientific ideas and criticism.

Collaboration-wide discussions and decisions in Fermilab 715, as in the vast majority of particle-physics experiments, were not limited to data and instrumentation, but also to administrative matters that could have bearings on members' careers. For example, the collaboration decided who could present collaboration results at which conferences:

> Our policy in this collaboration . . . is that an invitation to talk to this experiment is made to the collaboration and not to the individual. We don't care how it comes. There are times, obviously when an individual is invited, because they're looking at some individual for a job. And you don't send somebody else. But we reserved the right to make a collective decision as to who talks where. That's typically done in a group meeting.

As was also common in particle-physics experiments, manuscripts were reviewed and signed by all collaboration members before submission to a journal; and if published, the article included all members as co-authors. Fermilab 715 acted as a "total collaboration" in the sense that the breadth of the collaboration's jurisdiction extended into more areas than bureaucratic and semi-bureaucratic collaborations.

Not all particle-physics collaborations in our sample were participatory, and not all participatory collaborations were from particle physics. Occasionally, particle physicists incorporated bureaucratic features to ensure fairness and accountability when collaboration internalized different perspectives. The use of heavy-ion beams in particle accelerators has attracted the interest of physicists who had specialized in lower- and higher-energy experiments to investigate different phenomena, and one such collaboration in our sample incorporated bureaucratic features rather than relying on its members to work together well in a participatory manner. E814/E877 at Brookhaven, an international collaboration formed in 1986 for the investigation of electromagnetic and strong-force interactions in collisions of heavy ions, served to boost the sophistication of its physicists in heavy-ion physics before the commissioning of a new accelerator that would be dedicated to heavy-ion experiments.[28] The collaboration operated through matrix management. Subgroups of people managed safety, detector construction and maintenance, physics strategy, and analyses. The chair of the technical committee in charge of detector construction and maintenance could as easily have been designated "project manager," in view of his coordination and financial functions and powers. An officially designated spokesperson addressed scientific matters and external relations, while an Institute Board (with representatives from each of the participating organizations) dealt with the internal politics of the collaboration. Personnel issues were handled by the individual organizations but issues involving resources were handled hierarchically, with annual external evaluations that influenced funding.[29]

Occasionally, physicists in other specialties encountered circumstances that enabled them to organize participatory collaborations. The Sagittarius A experiment, a collaboration in very-long-baseline interferometry (VLBI), sought to locate more precisely the source of a distant emission in hopes of ascertaining whether the emission was coming from a distant quasar or a galactic center. Like Fermilab 715, this collaboration was built mostly on pre-existing relationships among astronomers and engineers with a shared goal—overcoming the technical challenge and achieving the scientific potential for conducting long-baseline interferometry at higher-frequency wavelengths. Like Fermilab 715, Sagittarius A participants could work within their ongoing funding arrangements and did not need formal contracts to satisfy fiscal accountability criteria. Like Fermilab 715, the appropriate division of labor within the collaboration was so obvious that no engineering leadership or hierarchy was required—the participating astronomer most familiar with each participating observatory assumed responsibility for taking data at that observatory. A scientific leader performed Sagittarius A's necessary coordinating functions like a spokesperson for particle physics; post-observation attempts to find interference effects in the playback of the data tapes and to interpret the interference patterns when they were found were collective efforts in which all participated equally. Where Sagittarius A most strongly differed from Fermilab 715 was that Sagittarius A required coordinated observing times from *several* independently managed observatories while Fermilab 715 required beamtime from *one* accelerator; Sagittarius A was an unusual project for the observatories while Fermilab 715 was a normal project for an accelerator laboratory.

Because participatory collaborations in other specialties are not typical, it seems justified to speak of "particle physics exceptionalism." The foundations of this exceptionalism reside in the idiosyncrasies of particle physics' institutional and engineering arrangements, not the character and culture of particle physicists. With funding programs in the United States dedicated to advancing knowledge of elementary particles, accelerator laboratories existing to support collaborative experiments, and with the experimenters themselves, especially in the case of fixed-target experiments, often capable of satisfying their engineering needs, particle physicists have routinely been able to collaborate only with those who share their cognitive ambitions. Physicists in other specialties also behave like particle physicists when they are able to duplicate particle physics' institutional and engineering arrangements. When particle physicists work in collaborations with multiple scientific interests, require an unusual facility or combination of facilities, or court exceptional engineering difficulties, they, too, introduce elements of bureaucratic collaborations.

Organization, Communication, and the Allocation of Credit

The last section of this chapter is devoted to describing the connections between these organizational forms, patterns of intra-collaboration communication and, the distribution of scientific credit in collaborative research. In the next chapter we take up the connection between organizational forms and technological practices.

Figure 3.2 provides a general description of three aspects of communication that characterize scientific collaborations. The largest difference is apparent between participatory collaborations and other forms. While all scientific projects involving multiple organizations hold meetings (whether face-to-face or electronic), some projects display more variability in the interaction among project members. For example, one set of collaborations has a period of intense communication punctuating longer stretches of less intense communication. That is to say, there is a "rhythm" or "temporal patterning" of communication—highly variable communication modes depending on the phase of the project, including a final phase of collective discussion, circulation, and signing off on papers.[30] This pattern is characteristic of participatory collaborations. These collaborations are also, as figure 3.2 shows, more likely to manage commu-

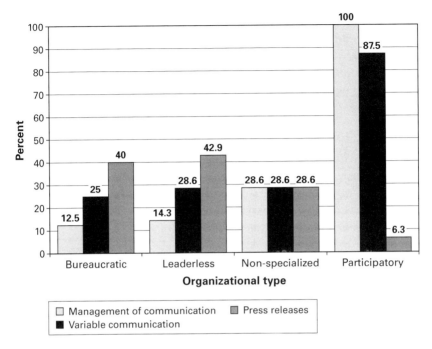

Figure 3.2
Organizational type by press releases, communication pattern, and management of external communication.

nication between members and outsiders and less likely to issue press releases.[31] Less dramatic but significant differences exist between leaderless and non-specialized (both semi-bureaucratic) collaborations. Non-specialized collaborations are more similar to participatory forms with respect to the management of external communication and press releases than they are to bureaucratic and the leaderless. To illustrate this disparity, consider the Keck observatory, used above to illustrate bureaucratic collaboration. Patterns of communication in a bureaucratic collaboration reflect the autonomy of member organizations in the context of a hierarchical internal organization. The objective of the collaboration, as we have seen, was the construction of twin telescopes. Organizational members designed an intra-collaboration hierarchy that required the instrumentation proposed by astronomers to be evaluated by the engineers. The latter were mightily concerned with staying on budget and on schedule; the collaboration hierarchy usually backed the engineers in disputes with astronomers who had risky ideas for improving the observatory. But the organizational members also formally limited the collaboration to the construction of the optical observatory and reserved to themselves the right to decide how their astronomers would use their observing time. Member organizations set up independent Time Allocation Committees to review proposals from their astronomers. With observatory use determined non-collaboratively, there was nothing comparable to data runs in particle physics that stimulated exceptionally intense periods of intra-collaboration communication.

With the Keck engineers empowered to stamp their views of a well-designed, well-built, and well-operating observatory on the collaboration, participating astronomers realized they could not influence the development of the observatory if they each focused on individual instrumentation concerns and did not regularize communication among themselves. Thus they matched the engineers in producing a steady flow of information and ideas for how the work should proceed. With stable patterns of activity throughout the duration of telescope-building, communication could be formalized, with two centers of incoming and outgoing information: the Project Office and the Science Steering Committee. The latter had regular monthly meetings and was, in the eyes of one participant, exceptionally formal:

> The science steering committee was probably a scientific committee unique in the history of astronomy, I would say. You don't usually get astronomers to sit down and run things but we met every month for many years and we had an agenda, we had people come to report to us all these instruments that were being built. Had a regular reporting cycle.

Once construction was completed and the observatory was in operation, the astronomers were left to their own devices to develop proposals for its use. Member organizations controlled the observing time and scientists competed for its use on terms of their choosing. Nothing formally prohibited astronomers from different organizations from working together on proposals and then sharing the observing time won from their member organizations. Yet nothing formally encouraged such collaboration, a situation quite unlike those projects that actively gave preference to collaborative work. Whether a successful proposal came from an individual, an intra-organization group, or a multi-organizational group, successful proposers became completely autonomous from the collaboration once they had taken data. The observers decided how to analyze their data, when and where to publish, and whom to list as authors without input from collaboration officials or other successful proposers. As a whole, the collaboration provided no support for communicating results to scientific communities.

The Keck project devoted considerable attention to public relations because the Keck Foundation was keenly interested in a highly visible project and because the UC system and Caltech recognized that they would benefit from publicity for having built the largest optical telescopes in the world. The collaboration took public relations seriously, and participating scientists appreciated professional competence in public relations:

Oh, I think the public eats it up. They love it. And X, who's been our public relations officer for the last few years, is very effective. He really gets out and gets news releases when interesting science discoveries are made. He really encourages photographers to come and magazine reporters and newspaper reporters to come so they can write glorious articles about it. If bigwigs want to have guided tours of the facilities . . . maybe they're on the board of trustees of Caltech and so Caltech obviously wants its rich guys to be excited about what Caltech is doing.

For a bureaucratic collaboration such as Keck, collaboration resources have been best spent on increasing awareness of the facility with the general public, while the career interests of participants are sufficient to generate news of developments within scientific communities.

Participatory collaborations exhibit precisely the opposite pattern of public relations and dissemination of results. In FNAL 715, for example, there were no press releases about the scientific or technical merits of the experiment. Its greatest source of publicity, the defection of a Soviet physicist, was unsought and caused consternation because it threw the collaboration's future into doubt:

[The defection] was in the headlines, I mean it was news. . . . It was at a time where that could have been a reason for breaking off the whole collaboration, everything. And it turned out that at that time neither the U.S. nor the Soviet Union chose to make it a major incident. The way it was portrayed is that he had done this for personal reasons. Which more or less was true, but not quite.

Communication was variable, depending on the phase of the project. While preparing instrumentation, participants only had to communicate to coordinate schedules and to ensure that the characteristics of their individual instruments stayed within readily defined engineering parameters. However, during the experimental runs, participants communicated on a daily basis. During the analysis of the data, communication again became sporadic—intense when a participant had a significant finding for the rest of the collaboration to review but otherwise infrequent. Rather than seeking public interest, FNAL 715 imposed strict collaboration-wide controls on presentation of results to the scientific community. Although individual members, especially graduate students writing dissertations, were principally responsible for drafting articles, manuscripts were circulated among all the participants for approval and only then submitted for publication as a multi-author paper to a physics journal.

In more bureaucratic projects, the dissemination of results was not a matter for the collaboration as a whole. Astronomers who won observing time on Keck did not have collaboration members who designed the telescope or ancillary instrumentation review, vet, or rewrite scientific results. Even Voyager, which needed a collective data acquisition strategy for the instruments on the scan platform during fly-bys of the planets, more closely resembled Keck than Fermilab 715. Separate Voyager instrument teams published their findings in the appropriate journals usually without consulting each other. Even when a scientific journal approached the collaboration about producing a special issue, the contributions came from the individual teams. They were generally not circulated and discussed by a collaboration-wide meeting or publication committee. DND-CAT, a less hierarchical collaboration than Voyager, also considered dissemination a matter for the scientists that won time from their organizations to use the beamline rather than an issue for the collaboration as a whole.

An exception to this pattern is the semi-bureaucratic, non-specialized collaboration. The organizational teams in these collaborations did not collect distinctive data sets that could be used for different purposes than the data collected by other teams. Yet the teams each wanted public credit for the work put into collecting data. In VLBI collaborations, participants put substantial effort into preparing an observatory for VLBI (and then returning it to its normal set-up) and into searching for the data-tape playbacks that would properly correlate to generate interference patterns. As in particle-physics collaborations, paper drafters distributed manuscripts for review and criticism, and all participants were included on author lists of scientific publications, even though most did not further analyze the correlated data, unless they asked to be excluded.[32]

Conclusion

Scientific collaborations exhibit tremendous organizational variability even in a set of related physical sciences. Through cluster analysis, we reduced the observed variability in structure to four types: bureaucratic, leaderless, non-specialized, and participatory. The notion of bureaucracy, as classically adumbrated by Max Weber, itself combines a multitude of organizational aspects, but for scientific collaborations, bureaucracy's most important features are formalization, hierarchy, the structure of leadership, and the division of labor. The image of scientific collaborations as informal, free-wheeling formations with strong communitarian aspects is sometimes accurate (Krige 1996; Zabusky 1996; Knorr Cetina 1999). But generalizations about the essentially informal and collectivist organization of large projects in science are often based on an analysis of particle physics, the one specialty whose practitioners cannot proceed *except* in collaboration.

The informal, participatory organization of particle-physics experiments can be attributed in part to the need for experiments to be student-friendly for the specialty to thrive. But the specialty's centralized institutional politics and competitive pressures render bureaucracy unnecessary much of the time. Because collaborations must submit proposals to accelerator laboratories, participants must commit to an organizational structure that convinces laboratory administrators (and their advisory panels) that they are properly organized to produce what they promise. With respect for internal structure thus secured before any commitment of resources to the collaboration is made, collaboration administrators have not required formalized powers to maintain order, and they can afford to grant broad rights of participation to all members of the collaboration, from graduate students to senior faculty. So while individuals play a prominent part in the inception and formation of experiments, they are blended into the communitarian structure once the experiment is fully operational. The result is a "superorganism" (to use Knorr Cetina's biological metaphor) engaged in cooperative work that downgrades the role of the individual and emphasizes community mechanisms such as collective authorship and free circulation of data. Such Athenian-style democracy has produced publications rather than cacophony because competition for discoveries and for career-advancing recognition limits the collective tolerance for intra-collaboration dissent.

"Particle physics exceptionalism" refers to the finding that most large scientific projects do not mirror the structure of those in particle physics, but vary substantially across fields in terms of organizational and managerial arrangements. Only the participatory category of collaboration, dominated by particle physics, is field specific, while the membership of the other three types is cross-disciplinary. Since few projects from other areas of physics share common features with particle physics—and it is highly unlikely that collaborations in the biological and chemical sciences would be *more* similar than other areas within physics itself—their marked characteristics must be considered exceptional.

What matters most for the structuring of this paradigmatic field is the breadth of activities and manner of governance. This form of governance is—somewhat counter-intuitively—intimately related to this breadth of activities. As we have noted, these collaborations cover the gamut of activities required to produce scientific knowledge. Data streams are collectivized. Collective decisions govern which groups address which particular topics with what subset of data. Results must be approved and communicated with collective endorsement. Work and managerial arrangements that are less formal than in other varieties of collaboration are accompanied by stringent controls over data acquisition, analysis, and external communication of results—the activities most important for career and reputation. While the power—indeed, the presence—of managers is minimal, particle physicists are more dependent on the will of the collaboration in the generation and dissemination of scientific results. In other forms of collaboration, managers were employed to exercise discretionary powers to secure what could not be obtained individually—but scientists then worked as individually as possible with what the collaboration provided.

Hence there is a relationship between the scope and organization of collaborations. *The broader a collaboration's activities, the more egalitarian its structure and the more participatory its management.* In the large, resource-intensive collaborations characteristic of modern science, the "organismic," participatory model is employed where the collaboration covers every aspect of the scientific process from instrument building to the publication of results. Most collaborations are simply not this encompassing and are unlikely to be in the future. Although there is always some anticipated output of scientific knowledge, communicated through presentations and papers, instrument builders do not generally seek to control it. Those that envision a spacecraft to Mars do not worry much about the submission of manuscripts based on the data the instruments collect. Managers whose decisions are accepted as legitimate, even to the point of overruling the scientists whose activities are the collaboration's reason for being, have little concern—and sometimes little curiosity—about the scientific content of papers submitted for publication. Where the range of collaboration activities is limited, decision making takes more hierarchical forms and bureaucratic organization segments the work of research teams to allow discretion within the boundaries of task assignments.

What must be understood about large scientific collaborations is their essential constitution from independent organizational entities that commit voluntarily to a temporary alliance for the creation of knowledge. Collaborations of high breadth are not simply agreements between individuals that take a single project from beginning to end. The counterintuitive aspect of the association is that greater scope *fosters* participation rather than the opposite. "Necessary bureaucracy" is inherent in the demands of the complex organizations and technologies utilized in modern knowledge production. Communitarian forms are appropriate responses to a situation where opportunities for individual contributions outside of the collective framework are minor and the final product will include dozens or hundreds of participants. Where the scope is broad, highly bureaucratic governance would be experienced as an undue and unjust exercise of power. But where the scope of a project is restricted, participants experience bureaucratic governance as a necessary evil that ensures that the collaboration takes appropriate account of interests and considerations that members from independent organizations consider vital.

Leaving aside these variations in detail, the overall pattern that emerges is bureaucracy by consensus. Where the agreement to work together is both voluntary and temporary, arrangements must be made to balance interests. A collaboration, even one that restricts certain kinds of freedom, is a risk taken for a larger gain. Consensus precedes the (several varieties of) bureaucratic organization that characterize collaborations in space science, geophysics, oceanography, medical physics, and ground-based astronomy. The agreement to participate is viewed as a benefit by the scientists and engineers involved. It is recognized effortlessly that larger collaborations require more structure—a relationship we confirmed in the preceding chapter. The general purpose of the consensus is often given by a committee that coordinates the project and balances the interests of the several organizations. Where collaborations operate with such a committee at the apex, the organizations tend to function as a group of equals. The commitment to collaborate often implies that at the directorial level there are relatively egalitarian relationships between representatives of participating institutions with differences in rank, seniority, and reputation that are often inconsequential, and chairs that are often temporary.[33]

Such projects can be both more or less structured depending on one's perspective. What is unusual—at least from the standpoint of classical complex organizations—is that the higher levels may be less bureaucratic than the lower. To be recruited into a traditional work organization is to accept employment in a hierarchical work structure. Students of informal organization have long known that patterns of practice are often organized into smaller, consensual groupings at the lowest task levels. This is true of scientific collaborations as well. But to speak of the importance of consensus before hierarchy is to capture the ephemerality of multi-institutional collaboration as well as the shared experience of graduate training that generates expectations for the conduct of scientific research. All collaborations are consensual in that organizational members are not compelled to participate. They are, however, required to submit to whatever hierarchy the collaboration creates.

As multi-organizational collaborations become common in more scientific specialties with varied patrons, it is especially important that managers and policy makers recognize that collaborations in particle physics—the "paradigmatic" collaborative field—are idiosyncratic. Their organization, characterized by Knorr Cetina as a "post-traditional communitarian formation" (1998) and labeled here as "participatory," is but one of several major ways to manage collaborations. American particle physicists have enjoyed a relatively uniform infrastructure of funding agencies, accelerator laboratories, and university-department facilities.[34] During the period covered by this study, electronic detectors installed at accelerator laboratories were the dominant experimental technique.[35] Competition for time and space at accelerator laboratories, routinized institutional politics, and the limited range of experimental styles heightened the competition for making discoveries and for testing theories. These conditions imposed extraordinary discipline that pushed collaborators to adopt similar organizational structures. But by sampling a broader range of physics collaborations, we found a variety of more overtly formal structures have served scientists facing different conditions.

4
Technology and Interdependence

It turns out that nearly all of the really exciting discoveries in astronomy in [the twentieth] century have been a result of technological innovation. So there's no question that the instrumentation side of things is absolutely vitally important, and I think it's often the scientists who get the glory for having discovered something.

—*astronomer*

Without interdependence there is no reason to form a collaboration. Without autonomy there is no reason to join one. What is the source of this precarious balance? Technology, broadly defined as the set of instruments and practices that researchers employ in the acquisition and manipulation of information, is the foundation of interdependence. All the multi-organizational collaborations we studied used sophisticated instrumentation. Scientists themselves are keenly aware of the crucial role of machines for extending knowledge in their respective fields, but the interdependencies configured around and by technology shape collaborations in other ways as well. This intertwining of knowledge and material practice, a perennial theme of science and technology studies, is explored in the present chapter.

By far the most common rationale for collaborating in our sample is that single organizations find it impossible to develop, acquire, or deploy all the instrumentation needed for scientifically meaningful or advantageous research. Thus, the critical stage in a collaborative project is often equipment design and building. Disciplines that make use of particle accelerators and the astronomical sciences rely more heavily on the construction of special instruments than areas such as materials science and medical physics, yet all are dependent on special machinery and protocols. Persisting social relations are generated in the process of encountering and solving technological difficulties. Instrument specification and construction can also be a source of tension and conflict, as this chapter shows. Thus, even in the relatively narrow sense of instrument construction and utilization, there are grounds to believe technology is crucial to understanding the collaborative process.

We use the word "technology" broadly, not just for machines and material artifacts but also in reference to the ways collaborations acquire and manage information. In our sense, the concept includes not only the design and building of equipment but also the practices of data acquisition, manipulation, and analysis. It includes methods of topical differentiation and the cross-checking of results. Such practices powerfully shape the daily habits and relationships of collaborators. Technology, as always, is both a constraint and an opportunity.

The breadth of activities in which participating scientists find themselves mutually dependent varies greatly across collaborations. As we saw in the last chapter, particle physics represents one extreme, a "total" collaboration. Its practitioners are mutually dependent for instrumentation, data acquisition, and data processing. These collaborations usually subject the dissemination of results to project-wide management. At the other extreme, some collaborations—materials science serves as one example—have operated as coordinating entities. Such collaborations facilitate research almost as "mini-funding agencies" that favor scientists who choose to coordinate their work and leave the participants free to make their own coordinating arrangements. Our most fruitful avenue to understanding instrumentation and interdependence is to put the emphasis on relations among research teams. Sometimes team structures coincide with the organizational members of collaborations, while in other cases teams are topic-driven and cut across participating institutions. In both instances the coordination of tasks, methods, data acquisition, results, and publications distinguish collaborative projects in terms of the interdependence of their constituent units. This interdependence is indicative of the degree to which participants perceive themselves as members of a community, albeit a temporary and diverse one.

Our argument is that technological practices are important forces in shaping and constraining the spheres of interdependence and autonomy within collaborations. Collaborations vary widely in the burdens they assume for acquiring and manipulating instrumentation: instrumentation may be built from scratch or the project may rely on off-the-shelf products. Even in the former case, projects differ in terms of their emphasis on innovation, from the development of unprecedented instrumentation to incremental modifications to standard tools. Apart from the design and construction of instrumentation, collaborations varied in the degree their participants depended on each other for data collection and analyses. Temporal phases of collaborative projects are often marked by specific patterns of communication regarding the design of instruments or the management of data. The intensity of communication is often high during the construction phase and slows once instruments are fully operational. After new instruments are built, providing some novel or special window on the world, there are still questions about what to do with the pictures seen. Will data be analyzed collectively or in teams? Will the topics addressed be overlapping or independent? Will results produced by one team be checked by another?

In this chapter we address the question of technology and interdependence in collaborative work. We begin by showing the degree to which scientific collaborations are inextricably interwoven with technological matters. This entanglement is not simply a matter of whether the project builds an instrument. The qualitative materials in the first section raise a variety of issues broadly covered by the theme of technological practices governing the acquisition and manipulation of data. The diverse roles of technology in scientific collaborations are described, focusing on the interdependencies that take shape as choices and contingencies arise. In the three sections that follow, these technological practices are related to organizational features described in the last chapter. The major connection that emerges is between the structure of leadership and the character of interdependence—greater interdependence leads to decentralization of leadership and less formalization.

Instrumentation and Interdependence

The most obvious significance of technology in scientific collaborations is the degree to which they are focused on instrumentation proper—the machine is the reason for the collaboration. If there is a "conventional view" of the role of technology in science, it is that new instruments are developed for much the same reason that motivates a company to develop a faster computer: competitive advantage. A participant in a telescope-building collaboration made the point especially well:

At the beginning, when we said we were going to make a segmented mirror telescope, there were many people who said it just wouldn't work. In fact, they thought that it was laughably ridiculous . . . and then we went through this long period in which we were not doing a very good job of making mirrors. . . . So people could see that we were having trouble. We didn't try to disguise that so people laughed even more and then we went on-line and of course we weren't perfect on day one, but as time goes on, people are, I think, fearing this telescope more and more.... And I'm hearing cases in which people are just going to stop doing certain kinds of scientific work because they think they can't compete.

What initially was viewed as "failure" even by the participants themselves came to be seen as the successful construction of a device that *discouraged* other scientific teams from pursuing related projects.[1] But the design of new machines is far from the whole story. There are many reasons large projects for the production of knowledge are "technoscientific" collaborations. While some collaboration seeks state-of-the-art technology, others are more interested in coordination and the recombination of existing technology. Projects experience varying degrees of technical change over their course. The design and construction of machines that is the principal source of interdependency for many projects is not nearly as salient for others as the management of information. The phase of data collection appears much different to those collaborations that take data collectively than to those that do so autonomously. Finally, data quality and checking are paramount for some collaborations, but in others they are independently managed by member teams.[2]

Four general ways of acquiring instrumentation were apparent in the group of scientific areas analyzed here. First, instrumentation may be constructed by the collaboration, a condition that often results from the need for highly specialized instrumentation remote from the interests of manufacturing firms. Second, instrumentation may be borrowed by customizing extant instrumentation or acquired through inclusion of a scientist developing instrumentation independently; collaborations that operate as coordinating agencies increase the level of interdependence by selecting subprojects that make use of multiple facilities that are already in place. Subcontracting is a third method. In this case the construction of instrumentation is the result of a need for specialized tools that overlaps sufficiently with manufacturing interests; but there is variation in the extent to which the collaboration or the subcontractor does the bulk of the design work. Fourth, instrumentation may be directly purchased in cases where there is need for less specialized, more flexible instrumentation that the user can adapt to a variety of research opportunities.

What we initially did not appreciate—because particle physics was the first field we investigated and these physicists usually build one-of-a-kind detectors—was that the collaboration often does not share this emphasis on specialized, dedicated equipment. Other varieties of practice become important. About four-fifths of the collaborations analyzed here designed or built their own equipment. This interest and expertise in the toys of the trade distinguishes collaborations that are predominantly oriented toward the construction of a material device or development of a technique and those that are not:

> Some of our facilities here have to be not so much state of the art, but they have to be as commercial as possible and as turnkey as possible, because that's the nature of the experiment. I mean it's an analytical technique, and people are not terribly interested and terribly turned on by the intricacies of the technique.

> The diffraction data sort of see the average structure, but the real fine details of the distortion required NMR. So I'm hoping to give the impression that a material scientist doesn't necessarily need these facilities for a great deal of work. It's only part of his, if you like, armory of research techniques. So that's quite different from high-energy physics.

> [We] were all just totally entranced by the number ten [meter telescope], and by the notion that we would be twice as big as five. And so, actually it was an emotional commitment to a number which in the early days we didn't really have any technological basis for thinking we could actually meet.

The first collaboration relied on standardized instrumentation that was familiar to all specialists in the field; scientific novelty required deploying the instruments in a coordinated fashion. The second collaboration accessed and adapted an extant facility for its purposes when standard instruments were unable to satisfy criteria for scientific excellence. The purpose of the third was to develop and build an unprecedented instrument. The first illustrates collaboration that is uninterested in instrumentation; the second those that utilize instrumentation intensively but do not depend on a high degree of technical innovation, and the third, collaboration that challenges the state-of-the-art in measuring techniques (58 percent of the projects in our sample).

Collaborations that jointly design and build instruments usually employ a division of labor wherein teams located at different organizational sites build components that fit together. In return, teams acquire certain rights to use other components. In the quotation below, the interdependencies clearly surround the instrumental component of the collaboration but not the analytical phase:

> On the instrumentation there's not [much freedom]. Everything has to be approved pretty much, because everything fits together so intricately you can't have individual groups changing different parts without carefully evaluating the consequences with all the parts of the apparatus. But I think on the analysis and data sharing, that's much freer. People can certainly go off and do whatever analysis they want to do.

For collaborations seeking to use extant but scarce instrumentation in novel ways, interdependence was temporal:

It was very hard to schedule because all these antennas were being used for other purposes for radio astronomy, and so you had to find some time when they could all work together. And of course you'd try to say how about such and such a week, and then one observatory would say no because we are doing that.... So it was agreed at that point to form this... network.

Such collaboration does not entail much risk of design flaws but overcomes administrative obstacles to scientifically advantageous deployments of standard instrumentation.

Technology-based interdependencies, especially among participants accustomed to working independently, were a prime source of intra-collaboration conflicts. Systems engineering issues—where one team wished to change its instrumentation in a way that affected another team's instrumentation—were common challenges to collaborations that designed and built instrumentation. But changes in instrumentation during the course of a project, which occurred in 29 percent of our sample, even posed difficulties for medical-physics collaborations that test technologies with large numbers of patients rather than developing or designing new technologies:

The only problem is when you start a study you try to keep it the same throughout, you know, you can't just keep introducing new technology throughout... because you [have] a database that's pretty meaningless. And yet you're under quite a bit of time pressure because of this constantly evolving technology. And radiologists ... [are] real technical.... So if something's new they want it. And they think it's better. Just by definition if it's new it's better. We don't usually find that to be the case.

Even when collaborations used unchanging, standard instrumentation, ensuring that all the instruments were operated with care and competence posed managerial challenges:

Part of this network agreement was that the people affiliated with each observatory would carry out all the tasks needed at that telescope for everybody, not just for the experiment of their own scientific interest but for everybody else as well. ... That worked for a while, but then as things grew and the observing load became greater, it became boring, and people started not to be as careful as they should be. That's when a lot of mistakes started to happen.

Only when collaborations built instruments that were truly independent save for their common dependence on particular conditions or technology did the teams enjoy technical autonomy:

One or two instruments were never built. It was actually the original plan ... to build more end stations than were finally built, because it turned out that once the initial money for the project was obtained the financial climate . . .changed. Money became tighter. . . . There was another proposal that was made with another university group . . . and they never obtained the money, so that whole project was cancelled. . . . It had no impact on the collaboration.

The focus thus far has been on machines and physical devices. But important variability is concealed by an exclusive focus on the construction of instruments and the failure to distinguish integrated from autonomous data-collection practices. What is not rare—and in fact a virtual obsession for some collaborations—is the organization of information flow. As Harry Collins has shown in the case of gravitational wave detection (1998), one of the most revealing social processes in international collaboration comes in the aftermath of data collection, when teams begin to examine and check results, arguing phenomena, artifacts, and the meaning of data. Once data began to appear, where did they go? Who had access to them? How were they checked? Integrated data collection implies that the participants collectively sought information and therefore attempted to coordinate the parameters of the instrumentation or protocols used. Data were collected autonomously when participants did not take data together and did not consult each other on the use of instruments. Between these extremes were situations in which participants took data together but did not tightly coordinate the use of instruments and situations in which participants did not need to take data together but did prescribe how instruments were to be used. Issues of data management and analysis are technological to their core.

Collaborations that integrated data acquisition and shared all data streams across the collaboration usually found more challenge in managing data than collaborations in which teams collected data autonomously. Effective data sharing required protocols that enabled teams to use data taken by others easily and confidently. For the most part, the ideologies of organized skepticism (Merton 1973) and replication (Collins 1975) were part and parcel of the scientific collaborations we studied; the checking of results was the principal means by which they were instantiated:

With such a collaborative team, there's essentially no opportunity for anyone to do anything by themselves. It doesn't happen. . . . Along with an image you have a record of how that image was processed from the raw data. . . . Maybe I made a mistake in what I did. Maybe I screwed up somewhere in my algorithms or something, but you must be able to take the raw data and get to that image. Otherwise that image never went out of the house. That was one of the rules we established early on.

Occasionally, however, a collaboration failed to appreciate the importance of specifying the conditions under which data were available:

> One person [must] take responsibility for cleaning it up before sending it to other people to look at. And that was part of the conflict, I guess, in the angular correlation experiments. There was no clear structure on how data should be shared. . . . It seemed like there was some tension about when the data was shared, and I think it was the hardest thing on the students.

And for those collaborations that were comfortable with teams operating instruments in total autonomy from each other, such issues did not exist:

> At the present time we're operating in the mode in which ground based optical astronomy and private institutions [have] always operated, mainly, you take your data home and you can keep it for 30 years and never show it to anybody.

However, in our data set, these collaborations were the exception and not the rule. Three-fourths of the collaborations here engaged in cross-checking of results.

Sharing data does not itself imply anything about its analysis. Who "owns" the topics for the analyses that are the ultimate reasons for collaborating? The management of topics that collaborators will pursue with the information collected is often considered in the formative stage of a project. As we argued in chapter 1, this is one reason the formation of collaboration stamps an imprint on what follows. But such decisions are not always irrevocable. From the standpoint of individual investigators or teams, this is a question of analytical autonomy. From the standpoint of the collaboration as a whole, the issue is one of problem allocation—members must feel satisfied with their potential to reap rewards from their investment of resources into instrumentation.

In some collaborations that shared and merged the data streams of several teams, members feared there would be too few significant topics to be addressed with the data for the number of participants. Graduate students' needs for distinctive, career-building accomplishments were a frequent source of such worries:

> In particular we had to worry about the theses and graduate students, because you can't have the same thesis written by two different students. . . . [We] actually circulated lists of thesis topics with names on them. . . . You could not assign your student to one of those topics.

Others found that studiously neglecting such questions enabled original, senior members to reach enduring, informal understandings of who would address what:

In practice, people got interested in a specific problem or a specific course or a specific galaxy or a specific quasar or kind of quasar quite early, and it was recognized by the rest of the people—and more or less everybody had his little domain that he continued to pursue, for more years than he probably should have.

In contrast, collaborations with autonomous teams had to face the prospect of enabling the collection of data that too few people found significant:

There's a lot of pressure put on people to not take data and forget about it, to actually process it, so they would keep a record of who has gotten data before.... And if people are taking data and not processing it and publishing it then that's counted against them.

Even though collaborations are intentionally ephemeral, members must be conscious of maintaining a reputation that will make them attractive candidates for the next collaboration.

The extracts above disclose the importance of technology for the interdependence of scientific collaborations but betray that it is not a simple matter of instrumentation. First, there is variability in the extent to which instruments are the focus. Second, alternative ways of acquiring instrumentation characterize those that do exhibit such reliance. Third, coordination or reusing old instruments may be more salient than design and construction of new ones. Fourth, collaborations differ in their interest in technological innovation. Fifth, the exigencies of development often generate technological changes during the course of a project. Sixth, the interdependencies of instrumentation that characterize some collaborations are distinct from those of data analysis and management that characterize others. Finally, the allocation of topics and checking of results are subject to technological practices.

It should be clear from this summary that the interdependencies of technology are many, and a product of a variety of factors. They are not predetermined by the nature or complexity of the equipment required by many large scientific collaborations. In fact, not much can be learned about collaboration simply by knowing that, for example, the resources of several organizations are needed to build, say, a large telescope. The division of labor, the practices of data collection, sharing, and checking, even the fine structure of interdependence is not beyond the ability of an astute observer to capture. To describe precisely who needs what from whom is readily described for individual collaborations, as the above extracts show, but a more general characterization is required to get beyond a list of features and reflect on the interdependence of several fields. The interdependencies of technology are not a feature of the "world out there" but the result of social interaction, decision-making processes, and contingency. In the following sections, we build on the organizational analysis in the last chapter to explore the interdependencies of technology, focusing on the acquisition of instruments, the collection of data, the management of the analysis, and the checking of results.

Acquiring Instruments

As indicated at the beginning the chapter, all multi-organizational collaborations we studied used sophisticated instrumentation. Most do more than borrow extant instrumentation or fill out purchase orders to acquire the tools to collect data.[3] The organization of collaborations is related to their acquisition of instrumentation. This is most clearly seen by treating the design and construction process together, followed by the process of subcontracting.

Figure 4.1 illustrates the relationship of organizational type with the design and construction of instrumentation.[4] All three dimensions of instrumentation described in the legend were coded to reflect simply presence or absence of the dimension. The heights of the bars represent the relative proportions of cases that built their own instruments, designed their own instruments, or contracted out their construction. The figure demonstrates that leaderless projects—all of them—designed and constructed the instruments used in the collaboration, as did the majority of participatory and bureaucratic collaborations. Of the four principal types, only non-specialized collaborations tend to employ instrumentation that is at hand or readily purchased.[5] The relationship is explained by the peculiar scientific goal that was prominent among non-specialized collaborations. The scientific value of collaborations with an unspecialized division of labor depended on the creation of uniform, standardized data. A major virtue of self-designed instrumentation lies in the potential for customizing or improving data collection for the idiosyncratic interests and objectives of a project. A collaboration that aims to standardize the acquisition of information over a range of data-collection sites should not design instrumentation unless inadequate instrumentation exists for its purposes—a participant who produces an innovative design could well be making the collaboration's task more difficult.

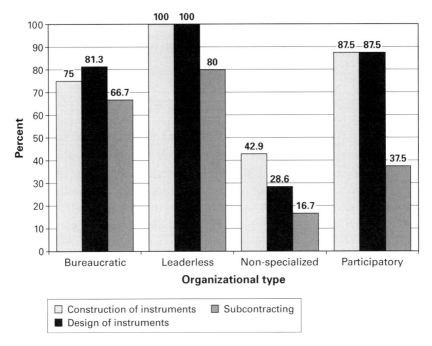

Figure 4.1
Organizational type by design of instruments, construction of instruments, and subcontracting.

It may be recalled that the International Satellite Cloud Climatology Project (ISCCP) sought to assemble a continuous record of global cloud coverage and cloud characteristics. Data from several satellites were needed to produce global coverage. The data from the several satellites required calibration against a common standard for the data to be internally consistent. In principle, ISCCP could have been a project to design and build a suite of satellites for cloud climatology. But even assuming the climatologists could generate the resources to carry out such a project, years would have passed from project inception to the beginning of data acquisition. Instead ISCCP took the data available from extant weather satellites—already standardized to meet the needs of meteorologists—and used several data stations to sample, calibrate, and process the information to meet the needs of a global processing center. As the extant satellites failed with age and were replaced by new ones, ISCCP relied on continuity in the evolution of weather satellites to ensure coherence in its long-term data-accumulation efforts. ISCCP was not an appropriate context for experimenting with novel ways to ascertain cloud characteristics that were customarily measured—such innovation could undermine the common calibration of the satellites. ISCCP was not the appropriate context for trying out measurements of novel cloud characteristics, because that could undermine the global coverage. A centralized hierarchy under a scientific leader served its need for setting standards of data collection.

Other forms of organization—bureaucratic, leaderless, participatory—supported a specialized division of labor enabling the design and construction of instrumentation. Allowing multiple teams to design instrumentation is a prescription for problems of compatibility and integration. In bureaucratic collaborations, organizational members provided leaders with sufficient powers to compel the production of mutually compatible designs. For instance, the Keck Observatory scientists who developed the novel mirror with elaborate interfaces to the rest of the telescope had two choices. Either they accepted the decisions of the project manager regarding the capabilities of the instrumentation the mirror would rely on or they appealed to the board of directors to compel him to be more ambitious in designing the interfacing instrumentation.

In leaderless collaborations, participants assumed that instrument designers had sufficiently generic and routine needs that no strong hierarchy was needed to keep these designs from interfering with each other. For example, the DND-CAT staff concentrated on the instrumentation that would deliver radiation with desired characteristics to targets, scientists at member organizations concentrated on designing end-station instrumentation that would detect radiation-target interactions, and the end-station and beamline instruments were expected to work together because end-station instrumentation would need only standard infrastructure support (space, power supplies, etc.) from the DND-CAT staff. Members of participatory collaborations assumed they would be collectively alert and dedicated enough to catch compatibility problems on the spot. In Fermilab 715, for instance, participants frequently checked instrumentation when the beam was off and averted disaster when one member found that the controls for a magnet were not properly regulating current flow through the magnet.

Subcontracting

All collaborative projects struggled in varying degrees with their desire to foster or support improvements in instrumentation while regulating developments in instrumentation to ensure compatibility and coherence

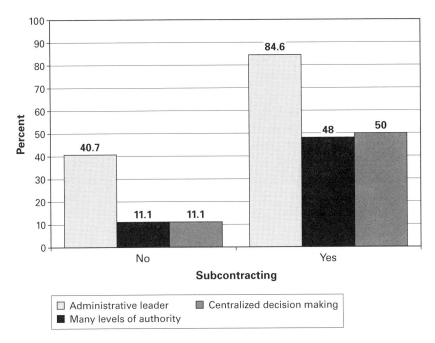

Figure 4.2
Subcontracting by administrative leader, levels of authority, and centralization of decision making.

within the proposed suite. Subcontracting represents another method of acquiring these facilities, conceptually independent from design and construction by the primary collaborating organizations. Subcontracting, a term that explicitly evokes the business of procurement, is a managerial concept with a sociological interpretation. It is the process of expansion or inclusion of new actors into the ephemeral cluster of collaborators. Yet these actors are not collaborators. They do not help to initiate the project and will not share in the recognition that will accrue if it is successful, except insofar as their participation helps to acquire future contracts. The assumption that collaborating organizations are interested in the *science* is not warranted—though it could be true—of subcontractors. Their motivation for participation is business, not knowledge. This characterization of subcontractors underscores both their peripheral status and the reason for their importance. Insofar as capabilities and commitment to complete necessary instrumental components are not available within the collaborating organizations themselves, outsiders are needed. Insofar as outsiders are needed, their activities must be brought under the control of the collaborators.

Subcontracting deserves special attention because of its association with organization. Participatory forms were distinctive in figure 4.1 in their avoidance of subcontracting. Figure 4.2 focuses on three organizational aspects associated with subcontracting: (1) the presence of an administrative leader or leaders (having one or more leaders whose tasks were mainly organizational and managerial), (2) multiple levels of authority (operationalized as having more hierarchical levels than a typical university department), and (3) centralized decision making (the degree to which leaders made decisions as opposed to the collaboration as a whole).[6] Collaborative projects elaborate formal organization when they expand their boundaries to incorporate external participants in the acquisition of technology.

Why do participatory and non-specialized collaborations avoid subcontracting, while bureaucratic and leaderless collaborations use it? Collaborations require little organizational structure when they divide labor for acquiring instrumentation along organizational lines *and* include organizational members with in-house facilities that are adequate to their tasks. In contrast, scientists in collaborations that required significant outside subcontracting for instrumental design or construction accept that a more bureaucratic organizational structure is necessary to negotiate with and supervise the external organizations. This conclusion clearly emerges from a comparison of a participatory particle-physics experiment with a pair of more bureaucratic collaborations.[7]

First consider Fermilab 715, our exemplar of a participatory collaboration. It had no administrative or engineering leader and did not subcontract for instrumentation. Most of its organizational members had been effectively functioning as ongoing businesses in the production of components for hyperon physics. The Fermilab 715 collaboration built on members' earlier activities. The Yale, Fermilab, and Leningrad teams all refurbished components they had previously built—or built a new version of what they had previously built, and all went on to recapitulate their 715 work after the completion of 715. The spokesperson for Fermilab 715 was explicit about his preference for using what the collaboration's members already had:

We don't invent new things in terms of apparatus. If it wasn't a piece of technology that wasn't readily doable we weren't interested. The Yale chambers from the earlier experiment were in fact very high technology when they were built in 1971. By 1981 they were not, but thank you we had them. Our motto is "Steal first." Plagiarism is the sincerest form of flattery.

Rather than create organizational hierarchy and limit the range of its members' participation in collaboration affairs, Fermilab 715 fostered self-sufficiency among its members, even to the point of foregoing the pursuit of technological innovations.

Examples of bureaucratic and leaderless collaborations from chapter 3—Keck and DND-CAT—provide clear instances of the connection between organizational formalization and centralization and subcontracting to acquire instrumentation. Bureaucratic and leaderless collaborations specified in legal agreements a schedule of payments through which members funded the collaboration. Both designated an individual to be its administrative or engineering leader.[8] In several cases, this individual's career was in project management. When effectiveness in acquiring hardware and keeping to budget and schedule was a defining element in the expected contribution of a project manager, subcontracting was more likely to be used to acquire instrumentation. In the case of Keck, an engineer served as project manager. Since organizational members formed Keck to build and operate one observatory on a finite budget— not an ongoing observatory development business—the project manager assembled a modest staff to let and oversee contracts for most elements of the observatory, including a contract with LBL to provide the segmented mirror.

In the case of DND-CAT, the staff director was a scientist who had previously helped to design and build more specialized synchrotron-radiation beamlines. This individual described himself as "enjoying playing with gadgets." Together with the chief staff members, "senior scientists slash engineers," a major structural grouping was created. Construction of the beamline was their full-time job, and getting the beamline quickly installed and reliably operating was their main professional challenge. To that end, said the staff director, "What can be purchased is purchased. We contract for the services of a small engineering firm that has experience with designing instrumentation like ours, so they design some of our components and supervise their construction."

Both Keck and DND-CAT are unusual in that they were once-in-a-career opportunities for their instigators to produce enduring facilities for their member organizations. But the relationship between organization and instrument acquisition is equally obvious in bureaucratic collaborations that were part of ongoing programs. The Voyager project, one of NASA's missions to investigate the outer planets of the solar system, likewise combined formal, hierarchical management with extensive subcontracting. A formalized organization was a foregone conclusion given the plans to include multiple scientific instruments, each of which would have its own set of co-investigators and make demands on spacecraft resources. All participants were bound together by contracts, even when the relationships were legally awkward, as was the case with an instrument proposal from the Goddard Space Flight Center:

> . . . based on our proposal there is a contractual arrangement which is even more complicated, because NASA has a contract with Caltech, and Caltech has a contract with [the Jet Propulsion Laboratory], and JPL has a contract with the accepted proposals. So we were really under contract to JPL. . . . But since this was legally not possible, we got our money directly from headquarters—but only after the project manager at JPL approves it or agrees to it.

A hierarchical organization was also necessary given the scrutiny to which Voyager would be subjected. Like all large NASA missions, Voyager needed approval from the Office of Management and Budget, and then from the Congress, before NASA began collecting proposals, so officials at NASA Headquarters were sure to monitor the project closely. Unlike most planetary-science missions, Voyager was able to rendezvous with several outer planets because of a rare alignment of their orbits, so that virtually all planetary scientists were interested in the project's capabilities and data. There was no question but that Voyager's management would be entrusted to a professional project manager at the Jet Propulsion Laboratory, closely supervised by NASA Headquarters, and that a project scientist would track engineering developments and argue the advantages of making the spacecraft as "science-friendly" as technically possible. For a project scientist, JPL broke with its customs and recruited a prestigious (though local) outsider, Ed Stone, a physicist from the California Institute of Technology. This was to ensure that communication between participating scientists and project engineers went through appropriate channels. The unsubtle subtext to this organizational structure was that nobody was beyond being disciplined through administrative channels and nobody could justify back-channel communication as a tactic for acquiring resources.

With a bureaucratic organization and an absolute deadline,[9] pragmatism in the acquisition of instrumentation was the order of the day. Subcontracting was frequently pragmatic for both engineers in project management and scientists responsible for particular instruments. Voyager's instrumentation, once launched, was beyond mechanical intervention and reliance on industrial experience in meeting standards of reliability and quality-control made intuitive sense. JPL contracted out most of the engineering subsystems to industry, reserving for itself only those functions and subsystems that could not be captured in contractual language:

> Most of the subsystems and the instruments were procured through a competitive procurement process with industry—with the exceptions of the structure and the cabling and the thermal control, which are very highly coupled to every other subsystem. It's difficult to write a clean set of requirements for those sorts of things, so we did them in-house. We get monthly reports from our contractors, and our people frequently travel to the contractor facilities and draft their own monthly reviews. That information has been the basis of our reports to Headquarters.

Some of the principal investigators responsible for specific instruments preferred to subcontract out the construction of the instrument and the details of some designs to industrial firms once they felt secure with their strategy for detecting and processing signals within the major engineering constraints (e.g., weight, power, telemetry) that JPL set for the project. The corporate subcontractor, rather than the scientist and his organization, then had the burden of staying abreast of the detailed standards JPL was imposing for the performance of materials and components that could be used in project instrumentation. Such was the case with the team under Rudolph Hanel of the Goddard Space Flight Center, which subcontracted to Texas Instruments for the construction of the infrared instrument for Voyager. Hanel and GSFC had previously set themselves and overcome the challenge of designing an infrared interferometer that could withstand the mechanical vibrations of rocket launch. He had successfully proposed the instrument for inner-planet missions, and had always contracted out construction of the flight instrument to Texas Instruments, which was in the business of providing clients with appropriate and reliable electronic circuitry. To propose and then build a Voyager instrument, Hanel worked on developing interferometry components suitable to the outer planets and relied on Texas Instruments and JPL to see that the electronics were suitable. These collaborators could thus concentrate on overcoming the idiosyncratic mechanical challenges posed by an infrared interferometer and on planning how best to use the interferometer, relying on the appointed experts at JPL and Texas Instruments to produce circuitry and shielding for the Jovian atmosphere.

Getting and Sharing Data

Distributing the burdens of instrumentation is probably the greatest challenge to keeping collaboration members happy in their work. The deployment of instrumentation must be geared toward the production of data

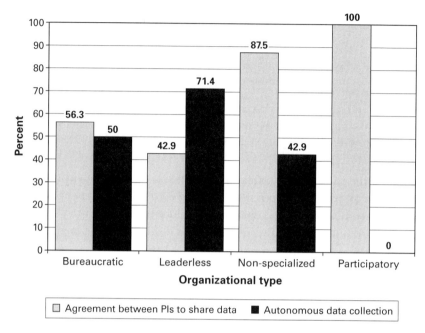

Figure 4.3
Organizational type by data collection and agreement between principal investigators to share data.

and, ultimately, knowledge, but there is considerable variation in how collaborations move from instruments to data. Some collaboration operates with a cultural norm of working together intimately to acquire and share data. Others produce ground rules to achieve limited integration in the use of instrumentation. Still other projects maintained the autonomy of the groups that designed (or acquired permission to use) certain parts of the instrumentation or the entire suite during particular periods of time.

Figure 4.3 depicts differences in the ways major forms of collaboration approach data collection and agreements to share data.[10] The general message is that more bureaucratic collaborations collect data in less collective fashion, with less sharing of information than in more participatory projects. But the relationship is more complex than this simple observation suggests. Hence, specific factors associated with a cooperative or integrated approach to data collection are illustrated in figure 4.4.[11]

The existence of contracts is a principal element of formalization. Collaborations that take data collectively are less likely to operate under a formal contract or other legally enforceable document. Note from

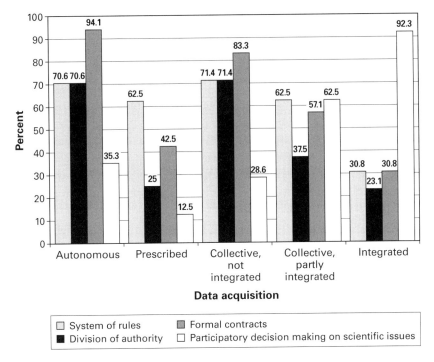

Figure 4.4
Collective data acquisition by system of rules, division of authority, formal contracts, and decision making on scientific issues.

figure 4.3 that leaderless collaborations, which have the highest rate of autonomous data acquisition, are the more formalized of the two semi-bureaucratic types. The implication is that collaborations, especially in the area of data acquisition, formalize their organizations in order to secure the autonomy of their constituent parts and limit the influence of the collaboration as a whole. The clearest contrast in data-acquisition practices is between the extremes of leaderless and participatory collaborations. Individual teams in the former disproportionately engaged in autonomous data gathering, while teams in all participatory projects took data collectively. Data-sharing agreements between principal investigators are patterned similarly, where the situation is reversed. All the participatory collaborations had such agreements; fewer than half of the leaderless collaborations did.[12]

DND-CAT, for instance, had six scientific leaders instead of one: five for each of the interdisciplinary teams of researchers who designed the end-station instruments and one for the group that built the beamline for the scientists. The collaboration was relatively formalized with data collection accomplished by the individual teams. Member organizations specified in their charter the proportion of funding that each institution would provide for design and construction of the beamline and end-station detectors. While the details of the instrumentation were hammered out in discussions between the CAT staff building the beamline and the multi-organizational teams of specialists, the charter also specified that the three organizations controlled beamtime in proportion to their financial contributions. Separate teams were assembled according to the particular interests of participating scientists. As a rule they did not envisage collaborating with other teams and did not develop agreements to share data. There were differences in disciplinary orientations, research foci, and organizational interests and it was clearly understood that research teams within the various organizations would respect the proprietary interests of the corporate members, as DuPont's representative made clear:

It will be our intent not to exchange [data] or share [data] processing unless there is an agreement going into the experiment that there is a mutual interest in the data. If we make some discovery or do some experiment that is scientifically or extremely interesting, and it was our experiment uniquely, we will come forth with data and reports that are processed at the end and share that with our collaborators. But I would say there is no intent or interest in sharing the processing, unless there is an intent going into the experiment.

Northwestern University scientists understood that any time they spent helping DuPont or Dow scientists to take data would be independent of their responsibilities to Northwestern:

[Collaborating with DuPont or Dow Chemical] will be done outside the university. University faculty get one day a week for consulting. That's part of the package. But no, this [collaborating] is not done on university time. It's either done on our own time or on university-assigned time for consulting. The university and a company don't have common interests.

Within teams, however, there was full cooperation among members on most of the experiments. This was driven by both the common scientific interests and the need to share instrumentation costs (e.g., DuPont's and Dow's polymer experiments). A few experimental stations, such as DuPont's fiber-spinning apparatus, were funded and utilized by a single institution for proprietary reasons. Nevertheless, there was not typically a need to exchange data between teams. Thus DND's member organizations used formal agreements to ensure that each could pursue its interest independently in performing experiments with x-ray beams.

Collaboration-wide administration was limited to producing the beamline and the detectors they all wanted.

Highly integrated data collection occurs when the independent interests of member organizations are less meaningful. These collaborations are less likely to be highly formalized. Particle-physics experiments routinely coordinate the parameters of the instrumentation they employ to acquire data and then integrate the data streams from multiple components. Particle physicists commit themselves and their parent organizations to experiments with no more formalities than signing proposals and later, when an accelerator laboratory has so requested, memoranda of understanding that specify the division of labor that the collaborators have already devised. Rarely are their participants concerned with defining and protecting the interests of their parent organizations.[13] For example, when the spokesperson for the particle-physics collaboration SLAC SPEAR 32 resigned from the SLAC staff to take a faculty position at the California Institute of Technology, he remained spokesperson, Caltech became a member of the collaboration, and Caltech, without intra-collaboration debate, assumed responsibility for part of the instrumentation.

Even in Fermilab 715, where one measurement was considered the experiment's centerpiece and the spokesperson was coming up for tenure at his university, organizational interests did not intrude. As in much of high-energy physics, this experiment required information from several detector components, each of which had to be adjusted for sensitivity to the same range of phenomena, in order to increase the chance of obtaining statistically significant signals for the processes under investigation. Participants from all teams took turns recording events from the hyperon beam line according to a schedule compiled by the spokesperson who was egalitarian when the time came to "mass the troops":

> The way that was done, there was a shift schedule. We kept three people on shift. I made it, which made me the least popular man on the experiment. I probably had my prejudices, but everybody got shift. And I tried to give everybody night shift, whether they liked it or not. I also tried to accommodate certain realities, like people who were at the universities.

This integrated approach to taking data led the principal investigators to share data without debate. As is often the case with particle-physics experiments, Fermilab 715 created a single set of summary tapes from which everyone worked, though not everyone was equally interested in all data streams. One group took the lead in the analysis of a particular topic for which a particular component's data were more or less important. Nevertheless, each team was responsible for providing the other with its data and the information needed to build scientific arguments from them.[14]

Between leaderless and participatory forms are collaborations that pursued less than fully integrated data acquisition within a semi-formal structure. For example, Voyager utilized an elaborate structure of contracts among NASA Headquarters, the Jet Propulsion Laboratory, and participating teams that had independently proposed and built scientific instruments. But the design of the spacecraft called for bolting the pointing instruments to a single "scan platform" such that they were not independently operational. The teams did not use the system of contracts as a basis for maintaining their independence and adjudicating disputes. Instead, under the direction of the project scientist to whom team leaders granted decision-making authority, arguments were raised concerning the scientific value of preferred orientations for the scan platform. Elaborate scripts for the orientation of the platform as the spacecraft approached each planet were collectively produced.

The second Greenland Ice Sheet Project (GISP2) extracted a surface-to-bedrock ice core from the summit of the Greenland glacier and performed numerous measurements on the ice and its trapped air bubbles. It resembled Voyager in data acquisition but was less formally organized. Grants rather than contracts committed the teams in GISP2 to perform their proposed measurements. Instead of management by an established organization like JPL that specialized in the management of scientific projects, one of the GISP2 principal investigators received an extra grant to create a temporary Science Management Office within his university. This temporary structure was not associated with fully integrated data acquisition and the teams were inclined to maintain their independence since they had produced independent proposals. Much of the instrumentation was not subject to any collaboration-wide systems engineering. This independence was possible because the core was sectioned, with pieces sent to the home-organization laboratories of the principal investigators, and desirable because the principal investigators were at varied stages of their careers. The Science Management Office facilitated exchanges of data by setting up a central file server[15]: data streams that dated the ice or its occluded air bubbles were essential for all members interested in developing a time series for climatological measurements. But teams maintained their autonomy by exchanging processed, (not raw) data and many resisted an effort to mobilize a collaboration-wide assessment of a particular climatological event. Managers realized they could not set collaboration-wide criteria for determining the point at which teams would make their data available

What remains to be explained is the case in which teams did *not* jointly collect data but the collaboration specified what kinds of information should result from the process of acquisition.[16] In general the more integrated the data acquisition, the more likely decision making is to be participatory. Conversely, when data is to be consolidated for general use, but is collected by different teams in different locations, participatory strategies are not generally used. ISCCP's development reveals the reason for this pattern. Its strategy of prescribed data acquisition stemmed from its participants' desire to create global data sets of cloud characteristics out of the regional data sets collected by weather satellites. Scientists were unlikely to use a global data set unless they felt confident that differences in the ways the regional data were generated would not skew the global data.

In theory, ISCCP could have sought to build the confidence of prospective users by formalization through legal agreements that committed the handlers of regional data to conform to specified standards. In practice the collaboration spurned formalization in favor of hierarchy to generate confidence in the quality of the global data sets. Scientists at a central global processing center reviewed the processing done by the teams and produced global sets with the algorithm it had spearheaded for deriving cloud characteristics from radiance data. ISCCP did not need both formalization *and* hierarchy. Hierarchy seemed the superior strategy for generating confidence since improvements in data-processing capabilities and interruptions in data-collecting capabilities (as when a satellite failed) were certain to occur. Changing circumstances would have forced a collaboration that relied on formalization to revisit its charter or to write a constitution that both specified requirements and allowed for adaptation. ISCCP could generate and maintain the confidence of users so long as scientists in the global processing center took advantage of innovations and patched up problems—which is to say so long as they were alert, well-funded, and technically competent. The perceived advantages of hierarchy over formalization for collaborations like ISCCP explains why they appear to use contracts and participatory decision making less often than needed—and why we do not find a direct linear relationship between data-acquisition strategies and organizational features.

Managing and Checking the Analysis

As was argued above, collaborative teams consider their scientific problems to be a question of analytical autonomy, while from the standpoint of the collaboration as a whole this is a matter of problem allocation. The management of topics is of much greater concern in some collaborations than others, especially when it is felt that the number of participants is large relative to the number of significant scientific problems that can be addressed with the data. Moreover, the management of topics implies not only a concern with the management of the data on which analysis is based, but also a concern with those students and postdocs whose careers will be shaped in significant ways by the subjects with which they are associated during their time with the project. Where autonomy of teams is truly complete, there is small worry about the accuracy and interpretation of results by other teams.[17]

The relationships between social organization, instrumental integration, and data acquisition are also evident in the management and manipulation of data. Collaborations that operated according to more formal procedures to maintain team autonomy in instrumentation and data acquisition continued these practices through the analytical phase. More participatory projects that integrated the design of instrumentation and data acquisition across teams did the same for data analyses. Collaborations that employ more hierarchical methods to require teams to meet data-collection standards sometimes persist in their requirement that teams follow certain procedures in analysis. Or they might explicitly terminate jurisdiction at the point data sets are made available for analysis.

These generalizations are suggested by the following graphs. Figure 4.5 illustrates the relationships of analytical management with formalization, hierarchy, and centralization. Figure 4.6 shows the same relationships for the cross-checking of results. Collaborations that do not allocate problems but leave management to their structural components are *more* likely to possess certain bureaucratic characteristics than projects in which the analysis is managed by the collaboration as whole. Similarly, collaborations that do not examine or cross-check results from multiple teams and instruments are *more* likely to exhibit these bureaucratic features than those that do so.

The use of high levels of formalization to maintain team autonomy in scientific analyses[18] is most apparent in the DND-CAT collaboration:

> The collaboration doesn't get involved with the scientific objectives of the individual members, except with respect to trying to ensure that they're building a

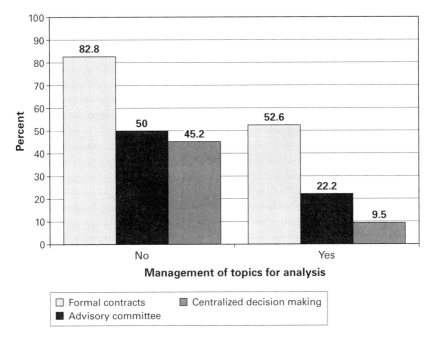

Figure 4.5
Management of topics for analysis by formal contracts, advisory committee, and centralization of decision making.

beam line that will meet our needs. I think certainly Dow and DuPont each have their approaches to doing data analysis. If we were to have the same data set in front of us we would probably pick different approaches.

To this DuPont scientist, an indirect benefit of the collaboration was the possibility that the proximity of researchers from different organizations at the beamline would enable the researchers to learn analytic tricks from each other. However, such learning was not an objective for the collaboration as a whole. It could most readily happen if researchers from all three organizations jointly performed an experiment from data acquisition through publication of results. But formidable organizational obstacles to such experiments were built into the collaboration. To perform such an experiment, the researchers would have to convince all three organizations to contribute beamtime and the time of the researchers, who could be doing proprietary research in the case of the corporate members.

Voyager was a highly formalized collaboration that partially integrated data acquisition to cope with the constraints of the scan platform.

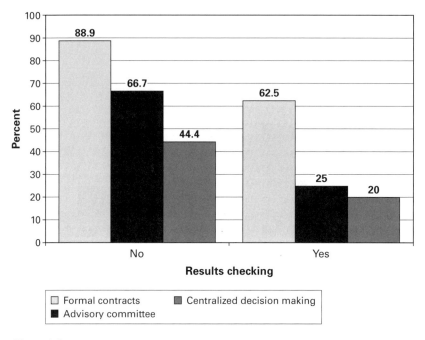

Figure 4.6
Results checking by formal contracts, advisory committee, and centralization of decision making.

The participants used formal procedures to secure the autonomy of teams in their use of data. Agreements between teams and project management provided for exclusive use of the data collected with their respective instruments for one year, after which the PIs were obliged to deposit their data in the Space Science Data Center located at Goddard for public use.[19] During the period of exclusive control of the data, teams focused on generating findings from the specific data streams from their instruments. A graduate student seeking to define a dissertation topic did not worry about the projects of researchers on other teams or clear his plans with the project scientist or any collaboration-wide official. All that mattered was establishing his relationships within the team:

Within the IRIS[20] team, I remember wondering myself what niche I could find for my own work, because it seemed to me that [other scientists] already had a commanding knowledge not only of the theory but also the . . . data. I suppose that there were times when we kind of collided with each other in our separate interests to do the same sorts of things. But I think that science sort of works that way.

Only when there was a compelling reason for combining data streams from different instruments did scientists seek inter-team cooperation—and that cooperation was not statutorily ensured:

> We're not *guaranteed* access to the other teams' data, so any access that we have is through largely informal arrangements with the other teams. . . . There were obvious collaborations that cried out to be done, for example, using our [infrared] data [from the instrument team] and imaging data [from the camera team].

The infrared data provided the basis for extracting temperature profiles; camera data were the basis for extracting wind patterns. A substantial body of theory and data related temperature and wind in the Earth's atmospheric dynamics. The leaders of the teams eventually worked out informal arrangements, but not with the speed and ease that would enable researchers from the two teams to cross-check each other's data and question the methods of each other's initial analyses before publication:

> I remember those of us who were interested in doing wind analysis on the IRIS team eagerly awaited the preliminary analysis and report of winds from Imaging. As a result of the Jupiter flyby, there was a detailed profile of the variation of wind with latitude. And when I first saw the profile, it seemed like magic to me that we had such detailed information, and yet we didn't have the numbers.

Such resentments were not the stuff of enduring conflict. Researchers on the IRIS and Imaging teams did find their way to cooperate in the use of each other's data. But the need for ad hoc arrangements and the awkwardness of establishing them are testament to the use of formal rules to guarantee the autonomy of the team leaders who invested time and effort in their respective instruments. Their natural inclination was to feel a proprietary claim to analyze the data first.

In contrast, analytical management was generally avoided by nonspecialized collaborations. ISCCP is a case where the leader declared scientific analyses beyond the purview of the collaboration proper. The justification for the collaboration was the existence of a community that desired to incorporate global statistics on clouds into climatological models by means of a global data set. The justification for an ISCCP global processing center was that one group of modelers should oversee the production of a global data set to ensure that the several data streams were competently merged and that the merged data set provided the information the modelers wanted. Had the global processing center done anything but make the data sets public, it would have undercut the justification for its existence:

From the beginning we thought in terms of the customers, meaning the modelers, not being the people doing the production [of the data set]. And without a documented existence of customers, you're not going to get the funding.... The problem is that if no modeler is involved, then there's no modeler perspective present.... The central processing is done by a modeling group—namely, GISS. Now, there's actually a precedent ... where model groups actually made data sets that they wanted and then also made them available to other model groups. And that's becoming much more common now.

As soon as GISS was confident that it had competently merged and processed the data the teams had produced by sampling and processing the daily streams from the several weather satellites, GISS shipped the data to the National Oceanic and Atmospheric Administration for archiving and distribution.[21] ISCCP participants did their best to dissociate the collaboration from the claims of the modelers who used the data:

People will sometimes read one of our data tapes, do something with the numbers and then call it an ISCCP result. [One scientist wanted to calculate an albedo from ISCCP values on reflectance.] So he did something that he thought was okay, and then he called it the ISCCP albedo in his paper. Fortunately, I was sent the manuscript to review, so I pointed out there was no such parameter on an ISCCP data tape. If he wanted to say he had read an ISCCP reflectance and that he had done something extra to it, that was okay. We don't mind taking the credit or the blame for the reflectance value, but don't blame us or credit us for something else you did. Clearly separate what you read from a data tape from what you've done.

Over time, misunderstandings of this sort diminished as individual scientists and journal editors became accustomed to incorporating ISCCP data into scientific arguments.

Teams in non-specialized collaborations did not lavish craftsmanship on developing unique instrumentation and consequently did not develop proprietary feelings for their data. As a result, these teams were not averse to checking each other's results or to authorizing an intra-collaboration authority to perform several analyses. In collaborations that prescribed data acquisition, data from a single team had little if any meaning until merged with the data from other teams. The same hierarchy that allowed these projects the means to prescribe data acquisition was also employed to check the data and to ensure the merged data met a standard of quality.

The scientific interests of teams in non-specialized collaborations were not tied to their instruments and the ways their instruments resembled or differed from those of other teams. Thus, the teams had no interest in analytical management, and the collaboration as a whole had little need to make policies in this regard—where autonomy in data collection was limited, freedom was provided to teams in data analysis. When one of these projects internalized many overlapping scientific interests, its scientific leaders declared that the collaboration was not in the business of performing data analyses on the merged data set—and thus threw open the opportunity to individual scientists whether or not they were collaboration members. When another could only address a narrow range of scientific topics, its scientific leader was required to determine who would perform data analyses.

Collaboration-wide checking of each team's data in these collaborations is abundantly exemplified in astronomical observations using very-long-baseline interferometry (VLBI). In this form of research, astronomers at widely separated radio observatories independently observe the same object at the same time and at the same wavelength. If the observatories are synchronized, pointed, and tuned with near-perfect precision, playing back the individual data tapes through a "correlator"[22] generates the interference pattern that results from the object's radiation having reached the observatories at slightly different times. The interference pattern can then be processed to produce an image with far higher angular resolution than can be obtained by any individual observatory. But if no two participating telescopes are synchronized, pointed, and properly tuned, participants have wasted their own time, wasted the time of the observatory staff, and used observing time that might have gone to successful single-observatory research. Ascertaining the quality of each other's observing at post-observation gatherings at a correlator site has been the central drama of these collaborations. The results were sufficiently uncertain that participants were loath to invest time in other data-processing tasks until correlation proved successful:

> You're playing [two tapes] together to search for this correlation, and you don't know where that correlation is because of the uncertainty in how the clocks are set up at the two [observing] stations. You try a certain alignment of the tapes; you cross multiply them, average for a long time and you look at the results. If you don't see any results then you shift the alignments of the tapes and look again. Sometimes you have to keep shifting to find something.

The drama of data checking has not been so high for other non-specialized collaborations, but the principle is similar. Whether the data sources are independent meteorological satellites that detect the same radiation wavelengths reflecting off clouds or physicians at several medical centers using designated techniques to diagnose diseases of the same organs, the collaborators cross-check data for compliance with standards that will make it possible to merge the data streams to form a meaningful data set.

The VLBI collaborations shunned management of scientific topics by being so tightly focused. They tended to be organized from the outset as one-topic, one-paper efforts that attempted to demonstrate the feasibility of VLBI at difficult-to-observe wavelengths. One of the senior scientists was the leader by virtue of having been the most entrepreneurial in bringing the participants together. For most other participants, the object being observed was more technically convenient than scientifically interesting. Once the collaboration had successfully correlated sufficient data streams to obtain interference patterns worthy of further processing, the considerable work involved in that processing was left to the leader, his students, and postdoctoral fellows who were seriously interested in the observed object. Other scientists participated in making technically sophisticated observations, and their reputations were correspondingly enhanced.

In participatory collaboration, including most particle-physics experiments, both analytical management and cross-checking of data from multiple components were common. Though a specialized division of labor was typical, with member organizations taking responsibility for different detector components, they merged the data streams from the components, since multiple measurements were necessary to characterize subnuclear interactions.[23] But particle-physics collaborations have not been so tightly focused as VLBI collaborations, nor have they labored to make their data widely accessible (as ISCCP did). Their data have usually been comprehensible only to those who mastered the intricacies of the instrumental design, and their data have usually supported the investigation of multiple particles or processes. Consequently, the management of analytical topics has been a lively issue, just as in more formalized collaborations. However, since specific teams in particle-physics collaborations routinely conceded the advantages of interdependence over autonomy, they have not generally claimed jurisdiction over management of topics, as did the teams in highly formalized collaborations such as Voyager. Instead the relevant actor has been the collaboration as a whole, with individual participants frequently claiming preeminence in the analysis of a particular topic—regardless of the number of authors included on a manuscript.

The members of FNAL 715 systematically pursued the checking of collaboration results. The center of analysis, as noted, was at Fermilab, where a crackerjack postdoc had a visiting appointment. But the spokesperson set up a separate data analysis program at Yale for the purpose of cross-checking the collaboration's claims about its preeminent topic: "[We] wanted to be able to check ourselves. We kept one [data analysis program] going at Yale. It was a short-handed effort. After it had served its purpose, we let it die." Because graduate students from the University of Chicago and from Yale University were members of the collaboration, and because the postdoc took the lead on the main topic, the collaboration was careful to manage other topics addressed with its data to ensure that other students had appropriately differentiated dissertations.

In SLAC SPEAR 32, the same ethos of collaboration-wide concern was evident both for the soundness of any individual participant's claims and for the differentiation of student dissertations without regard to the particular university in which the student was enrolled. Whenever a collaboration member presented an analysis at a collaboration meeting, "other people would check it. We didn't have a formal system, but it happened quite frequently. At these presentations, there'd be lots of discussion and lots of arguments. Somebody would go back and check something, or do something a different way, according to suggestions made at the meeting." Collaboration data proved relevant to a variety of physics topics, and members argued bitterly over the relative importance of topics when it was not feasible to collect data simultaneously. Still, when it came to the students and their individual interests, all claims for organizational boundaries and loyalty were abandoned, and what mattered was that the student found an appropriately differentiated topic:

There was no real distinction by institution. For example, I am at Caltech and I would be categorized as more interested in charm meson physics than in psi physics. Nonetheless two or three of my students did their theses on psi physics. People just thought of interesting problems and did them.

Conclusion

Chapters 3 and 4 introduced four types of collaboration, defined by their organization, and examined their relationships with the development of instrumentation, acquisition of data, and management of topics for analysis. These bureaucratic, leaderless, non-specialized, and participatory structures are shaped by interdependencies related to the instrumentation that is the primary rationale for most collaborative endeavors. Our premise is that decisions involving the acquisition of the tools required for producing knowledge are interwoven with practices for the sharing, the management, and the use of data. If that is so, it should not be surprising that the organization of collaborations has an impact on the analysis of data, or that instrumentation is fundamental to management. But the interdependencies of technology are complex. Participatory collaborations, for instance, are indistinguishable from bureaucratic and leaderless forms in the design and construction of instruments. Only the analysis of specific cases can reveal how projects elaborate or avoid formal organizational patterns. The story that emerges here is that the use of unproblematic instrumentation—purchased or easily available—is a condition for the production of uniform, standardized data that may be readily shared in processed form. Such collaborations elaborate bureaucratic organization, but their members have greater autonomy in a more limited sphere.

This interpretation in terms of autonomy can be extended to the connection between the character of leadership and the degree of interdependence: integrated data-acquisition practices are possible where the independent interests of parent organizations are less meaningful and are consistent with low levels of formalization. The "particle physics exception" we have noted involves routine coordination of instrumental parameters followed by the integration of data streams from experimental components. These scientists, based on an exclusive history of collaborative work, commit themselves to projects with a memorandum. Attention to the definition and protection of organizational interests is of passing significance—far different than the formalized organizational and managerial patterns that develop where data analysis is left to the discretion of independent teams. Where instrumental and social boundaries are consistent, as in the segregated acquisition of instruments, and where scientific and social boundaries are clear—as in the absence of data sharing and overlapping topics, bureaucracy protects independent teams and the organizations to which they belong. Participatory forms are a special case of overlapping technical and social boundaries in which the integration of practices renders organizational interests irrelevant. What remains is to consider collaborations as experienced by their participants, to show how the organizational fractioning of project work can replicate forms of scientific work that may be viewed as effectively non-collaborative.

5

Trust, Conflict, and Performance

One may win and one may lose.
The sky is big. There's lots to do.
—*astronomer*

CERN's managers were worried. For all the laboratory's technical prowess, its reputation lagged behind those of older, more established laboratories in the United States. CERN needed a win. One of its best bets lay in a collaboration led by Carlo Rubbia, an arrogant and demanding physicist who believed that the next fundamental achievement in particle physics would be the discovery of the boson. Still, the administration didn't fully trust Rubbia. His work tended to be hasty. Many felt he played fast and loose with the data. So they set another collaboration to work on the same task at the same time. The initial analysis of both experiments' data indicated the presence of the sought-for signal, but scientists and laboratory managers expressed varying degrees of confidence in this interpretation. Rubbia became progressively more insistent that his experiment had made a definitive discovery that should be reported immediately in the scientific literature. CERN's managers backed Rubbia. The discovery claim stood up, as the collaborations collected more data and subjected them to more refined analyses. Rubbia eventually shared a Nobel Prize for his role in the development of the accelerator and the detector used in this discovery (Krige 1998).

When research leads to a Nobel Prize, it is widely considered successful. But the conditions under which this collaboration pursued its objectives indicated skepticism of Rubbia's integrity and reliability during a high-stakes competition for the discovery of a hypothesized particle. Do we say that the research succeeded *in spite of* the distrust—that Rubbia overcame an atmosphere of narrow-minded cautiousness to establish the legitimacy of his bold claim? Do we say that it succeeded *because of* the distrust—that the discovery was due to the atmosphere of high-minded rigor and skepticism kindled by the competing project? Or do we say that the success was *independent of* the distrust—that the suspicions of CERN's managers and Rubbia's competitors had no relation to project outcomes?

Our final substantive analysis deals with the social processes and outcomes that have been intimated throughout the four preceding chapters. Three primary concepts—trust, conflict, and performance—are introduced in the next section. In chapters 3 and 4, the organization of relations between structural elements, particularly through their interdependencies with technology, were shown to be of central significance. This also holds true for the performance of collaborations. We begin by asking whether trust is important to the performance of the interorganizational collaborations that characterize late-twentieth-century science. The first message is that trust, which has been shown to have a central role in historical studies of scientific change, may not be central to the functioning of collaborations.[1] There is little evidence in our data that high-trust collaborations are more successful than low-trust collaborations. Given this finding, is there any basis for holding that trust is important? Finally, if trust is not important for performance, what is?

The empirical section of the chapter begins with two findings: (1) that there is no relationship between trust and important dimensions of performance and (2) there is not significantly greater trust in projects formed on the basis of pre-existing relationships. In light of these results, is there any reason for holding that trust is important at all? This issue is addressed in the following section. The central reason trust is believed to be important is that *projects with higher levels of conflict have lower levels of trust.*

Collaborations often need the participation of distrusting individuals. Moreover, efficient collaborative work depends on dividing labor in ways that are not conducive to building trust among those without some history. Conflict is inherent in the very heterogeneity of collaborations, but it need not be devastating, nor need it prevent the achievement of collaboration goals. Most large projects are equipped to handle conflict through (1) the formal organizational features on which they depend for the acquisition of resources and (2) the cultural expectation that conflicts be openly addressed and resolved. In most complex scientific collaborations, a low level of formal organization would court failure. Few scientists despise bureaucracy or argumentation enough to jeopardize the success of their enterprise. Conflict in research collaboration is rooted in the coexistence of distinctive structural components (e.g., multiple research teams, scientists and managers) and is associated with their interactions. What proves more significant than trust for the understanding of scientific collaboration is the organization of interaction between these structural components.

The last section of the chapter deals with an obvious question: If trust is not important for performance, then what is? We analyze the structural conditions associated with performance, using accounts by participants of their successes and failures, both managerial and scientific. Collaborations seen as successful by participants are typically those that experience uncertainty in the acquisition of resources. Put simply, the commencement of a project may be as important for its success as its conclusion.

Trust, Conflict, and Performance in Science

Present-day scholarship views *trust* as a fundamental requirement in all systems of knowledge production, especially where actors must coordinate their efforts toward a common goal. The recognition of "trustworthy persons" is a necessary component in building research networks, as Steven Shapin argues in his 1994 study of the development of modern science. Since collaborations are complex social formations, it may be that trust acquires even greater importance than in an era of individualistic experimentation.[2] Trust would appear to be crucial for working confidently in awareness that success depends on researchers in other organizations adhering to consensual objectives, research practices, technical specifications, and project deadlines. For this reason, trust in both inter-personal and inter-organizational relations has constituted a focus of scholarly interest both for science and technology studies and for organizational analysis generally.[3] A history of interaction is thought to contribute to the establishment of high levels of trust, while high trust is viewed as facilitating efficient cooperation in alliances, joint ventures, partnerships, and other forms of collaboration (Ring and Van de Ven 1994; Gulati 1995). Social ties reinforce a culture of trust, which in turn stimulates cooperation in organizations (Alter and Hage 1993). For example, trust is thought to be a prerequisite for the successful foundation of research-and-development consortia, because it reduces the costs of negotiation and minimizes opportunistic behavior in exchange relations between individuals and firms (Browning et al. 1995).

Within science and technology studies, this argument for trust as a sorting and selection mechanism is powerfully expressed by Karin Knorr

Cetina in her studies of high-energy physics (1995, 1999). Human agents maintain control over a complex set of instruments by building significant interactions on trust:

> Collaborations emerge from core groups trusting each other, but participants are also selected by the "money" and "manpower" they can contribute. . . . Within collaborations, therefore, "trust" classifies participants not in terms of the money they bring to an experiment but rather in terms of what is known about them: whose work can one build upon, whose results are "believable," and who does one wish to "cooperate with," and, alternatively, who does one wish to avoid? (Knorr Cetina 1995: 131)

Discriminations based on trust sort people informally, superimpose themselves on the formal aspects of professional status, and generate the important distinction between experts and non-experts in relationships with objects. Knorr Cetina views the promise of future experiments as one source of trustworthy behavior and non-bureaucratic techniques of control.

The organizational complexity of collaboration offers the potential for *conflict*. In the formative stage, collaborations with unconventional beginnings and those that are brokered by external bodies experience conflicts among participating organizations. Dissension between scientists and project managers over design issues is a frequent occurrence. Leadership struggles occur between principal investigators. Changing demands regarding data and instrumentation during the course of a project lead to assorted frictions. Conflicts over instrument design and construction, over shared data, and over when results are ready for dissemination are common. Indeed, for all the discussion of conflict in prior chapters, the reader might be forgiven for assuming that scientific collaborations were especially fraught with contentious behavior.

Conflict, of course, may be both disruptive and stimulating to the social fabric.[4] Indeed, the positive functions of conflict and intellectual contention are firmly rooted in both the historical analysis of intellectual strife pioneered by Thomas Kuhn (1970) and the sociology of philosophies elaborated recently by Randall Collins (1998). In scientific collaborations, too, conflict need not be unhealthy. As scientific projects encompass multiple organizations, the potential for conflict grows because its sources are embedded in the interaction system: functional differentiation between subunits, staff heterogeneity, divisions between scientists and project managers, styles of supervision, forms of power, and reward systems.[5] Conflicts may be caused by competition for resources, dissatisfaction over the discharge of tasks, and claims for credit—all common organizational processes—as well as by differences in technical approaches and scientific interpretations.

"Performance" refers to the valued outcomes of science, the criteria by which projects are justified and evaluated.[6] Organizations—including collaborations—are "instruments of purpose" (March and Sutton 1997), and assessing the degree to which that purpose is realized seems both possible and necessary. Yet measuring scientific performance is notoriously difficult. Complications and complexities arise because "performance" can be defined in multiple ways—beginning with varied output indicators such as publications, citations, and patents—and because those who have sought to measure performance have exhibited varied motivations.[7] In this study we focus on evaluative discourse—the judgments rendered by participants. Even if there were an independent and aperspectival standpoint from which performance could be determined, socially constructed *perceptions* would still be crucial. Collaborations may be perceived as better or worse in many respects—to what extent they accomplish certain objectives, whether the objectives are completed on time and within budget, to what degree their results or their instruments are used by others within and outside the field, and so on. These issues preoccupy scientists for the simple reason that projects require substantial commitments of resources and personnel. Performance judgments affect the reputations of collaborators and their likelihood of acquiring further resources, as information about outcomes circulates within a broader network of significant actors.

The role of trust is examined in more detail in the next section. Is trust related to the existence of relationships between collaborators before the onset of a project? Do collaborators with high levels of trust view their projects as more successful? In the next section, three axes of conflict are distinguished. We propose that the inverse association between conflict and trust is part of the reason for the continuing belief in the importance of trust and examine the factors associated with conflict. In general, conflict is a function of the interdependencies between the primary structural elements in collaborations—project teams, researchers and management, scientists and engineers. The theme of interdependence is extended throughout the discussion of performance. In the conclusion, we suggest that bureaucratic organization segments scientific work and can impose a structure for interaction resembling work that is actually non-collaborative.

156   Chapter 5

Trust

To note the existence of trust, to observe that it is a precondition of scientific work (or, indeed, of any project requiring coordinated social activity), or even to recognize that trust is a constitutive component of certain scientific developments does not imply an association with any *specific* social process or outcome. But the literature on trust generally fails to distinguish between what might be called the "foundational" varieties of trust that are necessary for any cooperative social enterprise and the more complex varieties of trust that characterize modern scientific organizations. In the former sense, scientists trust that others have something to contribute:

> They understand the chemistry pretty well, too. I understand the neutron diffraction pretty well. . . . I think we all trust each other and value, not just trust but then value the input of that other person. Because you recognize that they know more about that than you do. That's the basis of any collaboration, I think, or a lot of 'em.

This basic level of trust is necessary for collaboration in general, within science and without. For that reason it warrants no special attention. One important and neglected implication of this point is that it follows that scientists "trust" many colleagues with whom they do *not* collaborate, even as they "trust" their collaborators.

In another sense, individuals in collaborations are viewed as more or less trustworthy:

> When . . . one of the fathers of [this form of experimentation] was on it, I think there was very little [trust], and he was very, very critical of results, and so I think it changed. When he passed away . . . there was a marked change in the sense of these problems and so I think the trust went up of other people.

The absence of such trust excludes, often in the formative stage, certain kinds of collaborative relations between bitter enemies and antagonistic competitors. The individual mentioned above was only one of several former competitors participating in this collaboration. It was common, in this group of projects, for former competitors to collaborate. Why? One of the most important reasons is that only those with high levels of mutual animosity can resist a "brokered" collaboration in which a funding agency offers resources for a joint project:

> When these two proposals were put together, that marriage was done not by us but by the foundation trying to get two talented groups of people together and have one larger effort developed. And sometimes in that initial stage those two groups were fighting for control. Ultimately it was resolved, but there was a struggle . . . to gain control of it. . . . In the end it worked out fine.

In cases where competitors are joined, trust—at least at the outset—need not be particularly high.

If collaborations can consist of former collaborators, former rivals, or some mixture of the two, an obvious question arises: Is it actually the case that trust is higher where there were pre-existing relationships between the collaborators?[8] To the contrary, in our sample of collaborations *no connection was evident between the formation of collaborations through pre-existing relations and the overall levels of trust.* Projects that made use of extant social ties did not have higher levels of trust than those that formed from groups brought together by mediators such as funding agencies. It may have been the case that any overall advantage in trust through previous working relations was lost over the duration of the project.[9] Or perhaps trust is a more complex issue.

The theoretical literature on trust offers a clue to this problem in the distinction often drawn between "trust" as a feature of inter-personal relationships and "confidence" as an orientation toward institutions such as government and the media.[10] At stake is not the particular concept of institution used, which depends on the theorist, but the idea that some kinds of "trust" are not simply inter-personal orientations.[11] In the context of scientific collaborations, one team may have a great deal of "trust" in the scientific abilities of other individuals without having much confidence in their participation as collaborators with the will to set aside competing loyalties and to use collaboration management to define goals and to resolve conflicts. Conversely, one may have a great deal of confidence in the cooperative motives of one's collaborators in the face of mounting distrust of their research abilities.

In short, it is a mistake to focus exclusively on foundational trust embedded in inter-personal ties. This feature of dyadic relations is often distinct from the collective trust between social formations.[12] The latter is particularly relevant for a transient organization such as a scientific collaboration, in which the whole has limited power to compel compliance—for example, because the constituent parts are relatively autonomous organizations in their own right. In modern scientific collaborations, the relationships of interest are often between structural components such as teams.

Among the collaborations studied here, those projects formed from pre-existing relationships did not have any overall advantage in terms of trust. Does this matter? Surely there are important consequences that follow from the possession of such an obvious advantage. The reason—often implicit—for the interest in trust is its assumed association with certain kinds of performance. Scientific endeavors with high levels of trust should be "better," in some sense, than those without. Trust allows certain practices and assumptions to be taken for granted ("efficiency") and allows participants to achieve their goals ("effectiveness") while projects that are rife with mistrust should encounter obstacles and greater risk of failure. In the following case, for instance, the degree of mistrust between teams was relatively high:

> We can talk about the degree to which the astronomers at C trusted the astronomers in U and vice versa. . . . We often suspected that the C astronomers had hidden agendas that we couldn't see because it was so obvious that for the good of the [project], viewing us as being in one big boat, we should do X but nevertheless they didn't see it that way. They wanted to do Y, so we got very suspicious and we would [ask] what kind of internal problem were they trying to solve with this kind of stance.

In this case, not only did the teams suspect each other of having hidden agendas, even to the extent of not promoting the best interests of the project; the degree of trust between project management and its university-based teams declined:

> The . . . interesting question is, to what extent did the science steering committee trust the instrument groups that were building the instruments. We started out trusting them but as time went on I think we trusted them less and less because in some cases, we could see there was a level of incompetence there and people would come to us and say that black was white and it was clearly not like that you know so, as time went on, I got . . . cynical about some groups.

Note the difference between the two statements, particularly the distinction drawn between motives and competence. In the first, "hidden agendas" worried scientists at other universities. In the second, difficulties with the production of instrumental components threatened the quality of the final product.

Differences in perceived motives and competence are not the only problems faced by collaborators, but they are often the most salient. Teams in the preceding project (which involved the construction of a telescope) were characterized as having low levels of trust. Moreover, there was a lack of trust between teams and project management. Yet this project was viewed by informants (including the individual who provided the quotations above) as very successful. That kind of inverse relation between trust and success is not typical, but in our sample of collaborations *there was no significant relationship between the degree of trust reported by our informants and their perceptions of the success of the project.* That is, projects that exhibited higher levels of trust were not, on average, viewed as more successful than those with lower levels of trust.[13] Moreover, trust in other researchers was not related to finishing the project on time and within budget. The reason may be that performance is simply dependent on other factors, an issue to be raised later in the chapter. Generally high levels of trust do not mean the project will be successful, and lack of trust is not incompatible with success. Note the following example, in which a new galactic feature was discovered:

> In the beginning when we were all observing for each other you had to have other people take your data for you, and so there certainly was a lot of trust there. And in later years that wasn't true, but the scientific trust . . . was still there. . . . When the [instrument] was built, one of the first scientific results . . . had to do with the particular feature in a particular radio galaxy. And when the people doing this mentioned it to others . . . they saw the same feature in their data and they agreed. It wasn't a joint publication, but the two papers were published simultaneously in the same journal at the same time. Nobody tried to rush it. Even though they both knew about each other nobody tried to rush into print or beat the other.

What is characterized as decreasing trust in certain practices of other teams coexists in the same collaboration with fair play in the matter of credit through publication. Did it matter to the overall success of the project that teams didn't trust each other to collect data, or that they trusted each other not to steal the credit for a discovery? We suggest that it did not. Here, teams developed an alternative solution to the problem of data collection—they hired technical specialists and did not utilize the scientific labor of other groups. Trust, to use the above distinction, became confidence, and resided, in the end, in data collected by alternative means. Would the project have been less successful if one team had rushed to publication? Such an argument would be difficult to sustain. However, such practices may well be associated with conflict.

In sum, where foundational trust is concerned it is uniformly the case that individual collaborators have this kind of trust in each other's willingness and ability to contribute to collaboration. Foundational trust excludes collaborations involving scientists who aggressively doubt each other's scruples or competence. However, trust as the complex process of ongoing relationships between teams is related neither to pre-existing relationships among collaborators nor to their assessments of a project's success. In the next section, we propose that the association between conflict and trust is one reason for the continuing belief in the importance of trust. Three axes of conflict are distinguished: between teams, between researchers and project management, and between scientists and engineers. Analysis of the structural sources of these conflicts shows that interdependence is often their root.

Conflict

An elementary potential for conflict is built into all social relationships. For many scientists, simple argument or disagreement is as likely to be interpreted as a favor as it is as a challenge and need not rise to level of conflict. Conflict is an issue for scientists who compete for resources and reputation—the state of play in both routine and controversial technical disciplines. Less often noted is the clear potential for conflict in complex scientific collaborations in which resources, technical approaches, and deadlines must continually be debated and resolved to meet a project's objectives. What emerges from this group of collaborations is that interpersonal difficulties may be evident, even pronounced, without affecting the collaboration as a whole. As such, these conflicts do not warrant extended analysis, however memorable they may be to the individuals that experience them. When such conflicts become obtrusive, other collaborators urge the parties to see their differences as irrelevant to project goals. Tasks may be redefined and redistributed to minimize interaction between disputants:

Not everything is separable, and it requires the ability to go and consult. . . . Where we haven't been able to [consult], the communications take place at the staff level. . . . I've had to intervene when one of the X programmers is constantly fighting with one of our programmers. . . . This went on for a couple of years and "Stop it, guys." You know, "Come on." . . . Programs are not just lines of code . . . and these two guys were, in terms of their actual personality and their programming philosophy, on opposite ends of the realm of what's possible in the universe. And they managed to work together.

Many collaborations experience constant small battles among individuals over the use of resources:

There's a project going on at X that I think wastes too much telescope time for the scientific results, and I'll bitch and complain. . . . It's sort of like what goes on in any family. You have a fight with your wife, it doesn't mean you get a divorce.

Where universities are involved in such conflicts, the exchange of personnel through the hiring of each other's students as postdocs and junior faculty members tends to work against long-term disputes between organizations.

But serious conflict can occur between teams or structural groupings that collaborate. Whether viewed as threatening or productive, constraining or liberating, such conflicts must be taken into account by participants. A scientist who had enjoyed smooth relations with an organization when he was a local user of their resources found matters very different when he sought to involve that same organization in a collaboration that his new employer would be leading:

> I would say there has always been a tension with X for reasons that I don't fully appreciate. . . . X is an organization that has very bright and strong individuals, and it's an organization that's kind of used to being in a leadership role in running projects. And so I think in general what I'm learning is that they don't partner well.

Note the absence of specific inter-personal difficulties in this comment. No individual participants are cited as having difficult personalities or idiosyncratic work styles. Note also the significance of the conflict to the collaboration's overall functioning. A corps of talented individuals were not giving the collaboration their best effort, and the interviewee seemed saddened by X's disaffection.

One of the most serious types of inter-group conflict reported by informants occurs between teams, often over such issues as resources, communication, and credit as well as control of the project. The specific sources of conflict vary widely from one project to another:

> Everyone is very far away from one another, and again you have a culture conflict in that X people don't use email—they don't regard email as an appropriate form of business communication. And at Y of course it's second nature. . . . So in the early history of the project we would pepper them with all kinds of mailing lists and informative updates and memos and reports electronically, which they either didn't read or they coded as not important. Because in the X world if it's important then it's on paper or it's a phone call or it's a face-to-face meeting, and it took us years to figure that out. And those were important years, when X was basically just clueless about what was happening.[14]

Once such conflict is embedded in inter-group relations it is difficult to eliminate. In the same project, difficulties did not end after both partners began to use email:

> [From] Y's point of view, X had made all kinds of promises that they couldn't keep. From X's point of view Y had sold them a bill of goods, which was we were all going to be in this together building this system, cooperating. . . . From X's point of view, Y's idea of cooperation was "We get all the power, and you do what we ask you to do when we ask you to do it."

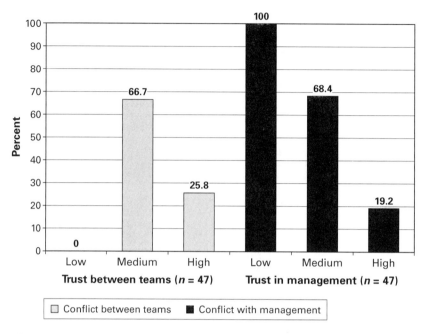

Figure 5.1
Relationship between trust and conflict.

The ownership and division of control over the project between the major organizations could not be resolved by the simple introduction of a new communications technology.[15] Figure 5.1 shows that in this sample of collaborations, trust and conflict are inversely associated.[16] The "low trust" category comprises only a few cases. However, two-thirds of projects with moderate levels of trust toward other teams experienced conflict, compared with only one-fourth of those with high levels of trust. This relationship applies to project management as well. Fewer than one-fifth of collaborations with high trust in management experienced conflict with management, compared with more than two-thirds of those with only moderate trust. Conflict is rare in projects with high levels of trust. If collaborations with lower trust have higher conflict, this serves as one part of the explanation for the perceived importance of trust.[17]

There are many possible explanations for the association between trust and conflict. It might be directional or causal, if trusting relations actually reduce the potential for conflict. Indeed, theoretical grounds have been adduced for supposing that trusting social relations promote the convergence of interests and a neutralization of conflict (Zaheer et al. 1998). But the causal relationship may be reversed if increased conflict reduces trust, just as friends attracted to the same romantic partner soon develop suspicions. What seems most likely is that trust and conflict are more intimately intertwined than is allowed by any simple causal model.[18] Trust both shapes and is shaped by the flow of negotiations, practices, and behaviors that can breed tacit disputes and overt conflicts. As perceived interests diverge, small differences arise in the course of interaction. Disputes may be anticipated before they actually occur—we do not know what proportion of conflicts is spontaneous, surprising participants with their rapid emergence. The line between expected conflict and diminished trust may be thin indeed.

Take the example of one telescope-building project. At the outset there was apparently a high level of trust between project teams. Asked about the degree of trust on this project relative to her experiences in academic departments, one informant laughed, replied "High—too high sometimes," and proceeded to discuss problems with timely progress owing to delayed delivery of instruments and disagreements over their features. The initially high level of trust was diminished, and the project ended in significant conflict. Trust had not prevented the conflict, and conflict eventually reduced trust.

In another instance, a European research team was brought into an existing collaboration. There was initially a deficit of trust, owing to the perception that this new group would use instrumentation already developed by one of the existing teams to pursue a science program that overlapped these pre-existing interests. Yet eventually the two teams were able to define sufficiently different research topics. When the expected internal competition did not arise, trust developed. One of the teams, viewing the experimental results of a collaborator, explicitly refrained from doing what it considered to be a better experiment because of the potential for "political problems" within the collaboration. Such decisions produced micro-climates of trust that reduced conflict. Though we are confident that trust and conflict are associated, the dynamic interplay between them does not admit of formulation in any simple way.

What, then, can be said about the factors associated with conflict itself? The primary lines of conflict in research collaborations are between major structural groups. Analyses of collaborations in high-energy physics, space science, and geophysics led to the identification of three significant varieties of inter-group conflict: conflict between teams, conflict between researchers and project management, and conflict between scientists and engineers.[19]

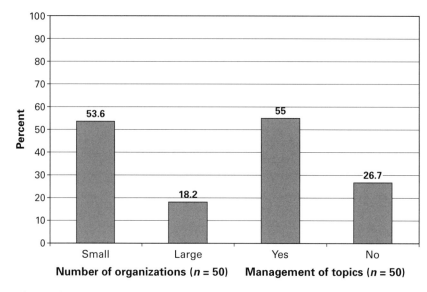

Figure 5.2
Relationship between team conflict, number of organizations, and management of topics.

Conflict between Project Teams

Figure 5.2 provides evidence of two factors associated with conflicts between teams[20]: collaborations with *fewer organizations* and collaborations that *manage the topics their members analyze* are prone to between-team conflicts. In the first instance, projects involving fewer organizations are more than twice as likely to spawn such conflicts. Smaller projects focus competition within the collaboration over responsibility for choice tasks (and avoidance of responsibility for undesirable tasks). In collaborations with fewer organizations, the stakes are higher for each. Large projects are visible and expensive and cannot afford to let internal conflicts endanger the mission.[21] Similarly, more than half of all collaborations that sought to manage scientific topics experienced team conflicts, versus only one-fourth of those that did not seek to manage scientific topics. The right to address particular scientific topics—and to take credit for addressing them successfully—is one of the most valued aspects of collaborative participation on projects in which great effort has gone into the acquisition of resources and instrumentation. Small collaborations that managed topics for analysis are the collaborations most likely to explode along the lines of their member teams.

The process is illustrated by the comparison of the Positron Diffraction and Microscopy collaboration at Brookhaven with the PEP 4-9 collaboration at the Stanford Linear Accelerator Center. The former was a materials-science collaboration that consisted of a handful of respectful competitors and a number of students and technicians. The latter was a large particle-physics collaboration that consisted of two independent collaborations, the larger of which included many independent physicists and a number of students and technicians. Both collaborations were developing unprecedented capabilities for taking data. In the case of the Positron Collaboration, Brookhaven's facilities afforded the potential for developing a positron beam with orders-of-magnitude higher intensity than a university could support. SLAC PEP 4-9 was developing a novel detector with unprecedented capabilities for tracking and identifying the charged particles that emanate from high-energy electron-positron annihilation.

Both collaborations internalized distinctive experimental styles and overlapping scientific interests and ambitions. Both collaborations, for somewhat different reasons, viewed collective management of instrumentation as appropriate. The handful of senior physicists in the Positron Collaboration considered consultation on topics an expression of mutual respect and a useful review of the quality of their plans. The many physicists in SLAC PEP 4-9 viewed collective management of the detector's use as a necessity to ensure that teams were able to reconstruct events using data from all the detector's components and to ensure that the teams were rewarded for the immense effort they were putting into hardware development. Consequently, both collaborations inspired worries about "who gets to do what" with their instrumentation.

Yet the trajectories of the two collaborations diverged radically when these worries became an object of action. In the case of the Positron Collaboration, a wrenching reorganization became necessary when a collaboration member used the instrumentation without consulting others in order to experiment with a new sample. The material was of such obvious scientific significance that no one would have disputed the value of the experiment. But to some the action was an unpardonable grab for individual scientific credit while to others it was an ill-conceived way to avoid dealing with the difficult personalities among the collaboration's senior physicists when no question of scientific values was at stake. All agreed, and, as one participant noted, "That was the end of the consortium. The consortium broke up after 3 years of funding by the NSF." The senior scientists also agreed that experimenting with intense positron beams remained too desirable not to pursue. After a hiatus during which Brookhaven rebuilt the positron beamline to remove it from the immediate environs of the high-flux beam reactor,[22] the collaboration reconstituted itself with a different organization and a different understanding of team relationships:

> When we became a consortium again, we were called a Participating Research Team. [Our organization is] better in that we are much more comfortable recognizing that some of the experiments associated with the PRT are really not in everyone's interest. Everyone does not pretend they're doing these experiments. We're not a consortium in that sense.

The collaboration remained small, and all members helped with the instrumentation that brought focused beams of positrons with a variety of characteristics onto a target. But teams were entitled to develop their own materials for use as targets and their own instrumentation to examine target/beam interactions.

In contrast, SLAC PEP 4-9 was able to quash intra-collaboration worries about who could do what with its detector before anyone acted on such fears. As its members struggled to work through the difficulties of bringing its innovative detector into existence, other collaborations using less ambitious detectors began publishing papers on the processes and particles being produced by the PEP accelerator. Like the Positron Collaboration member who feared he would not be able to perform the measurement he had developed the means to pursue, PEP 4-9 teams began feeling urges to claim physics topics as their own lest they be stuck working on something that the efforts of other collaborations might have already rendered mundane. Unlike the Positron Collaboration, the members of the PEP 4-9 were uninhibited in collaboration-wide discussions—no single, difficult personality would make a difference. PEP 4-9's response to worries about how physics topics would be managed was to hold a retreat:

> We were a collaboration that had to put resources into hardware . . . so we couldn't turn data into physics as quickly as we might. . . . In that interim [when the collaboration had the hardware working but not yet produced the software for data analysis], we decided to have a collaboration retreat. . . . And I remember we went into it with a group of about 25 very concerned graduate students, because it wasn't clear that there was enough physics for 25 theses. We broke into groups of about 30 people to discuss four different physics areas. And then we would come back and report to each other. It was absolutely clear by the end of the weekend that we had of the order of 50 theses of data and all the graduate students were feeling well fed.

In the small collaboration with collectively managed instrumentation, one person's eagerness to be the first to investigate an important topic produced collaboration-threatening conflict. The larger collaboration with collectively managed instrumentation defused the potential for such conflicts by providing its members with ample room for intellectual maneuver in the definition and choice of topics to address.

Conflict with Project Management

The second major axis of conflict in inter-organizational collaborations occurs between participating scientists and collaboration management. Of course, the potential for such conflict can only exist in collaborations that distinguish project managers from scientific personnel. We found such conflict had disciplinary roots but is also associated with organizational features. The key element in the differentiation of types of project management is whether participating scientists felt that the collaboration's budget was controlled by a non-scientist with independent professional standing. For example, note the distinction this participant drew between scientists and managers:

[The project manager] always controlled the budget and the thing I learned from this project is that money is power. And so it would happen occasionally that the scientists would make requests, more than requests, and they would not be honored, and in general, there was an atmosphere of tension there.

For scientists to prevail, they felt obliged to go over the head of the project manager and aggressively argue an intellectually rigorous case to higher authorities:

[Project management] wanted to move forward with what we knew at that time.... The astronomers said the quality . . . isn't good enough. We—me actually—documented this case with simulated observations estimating what the image quality would be if we went ahead and used this technique and we made a presentation in front of the board suggesting that this would be a crummy telescope and it would be a laughing stock, and Mr. [donor] wouldn't like having his name attached. . . . We wanted the board . . . to postpone construction . . . and they agreed.

In this case, this conflict between management and scientists was refereed by the project board owing to persistent disagreements.

Figures 5.3 and 5.4 illustrate the factors most closely related to conflict between scientists and project management. The first figure shows that *collaborations in the field sciences—space science and geophysics—were more likely than laboratory sciences to spawn conflicts between scientists and project managers.* Field sciences are logistically more challenging than laboratory

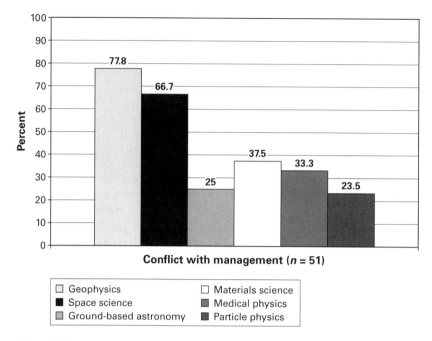

Figure 5.3
Relationship between field and conflict with management.

sciences, and the logistics of field work are relevant to commercial, military, and intelligence interests. These larger interests have given rise to professionals specializing in the operation of instrumentation in hostile environments. When scientists formed collaborations to perform large-scale field research, they generally found it necessary to work with such professionals, either as a matter of a funding agency's policy or as a result of their own sense of which roles to include in the organizational structure.

For example, the lone American academic geophysicist in the first Greenland Ice Sheet Project (GISP-I)[23] had long worked for the Army Corps of Engineers' Cold Regions Research and Engineering Laboratory (CRREL), where research and development of ice drills for glaciological and climatological research was an obvious offshoot of the army's interest in improving the civil engineering needed for military operations in the Arctic. GISP-1 drilling was both greatly assisted and limited by the researchers' need for support from the military radar bases maintained by the United States on Greenland. When GISP's results inspired broad demands among American academics for an ice-coring project that best

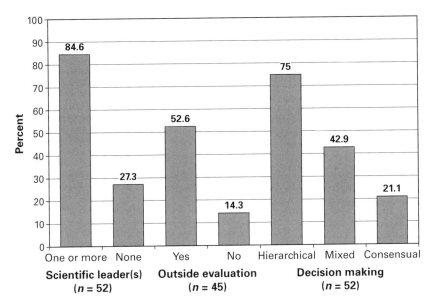

Figure 5.4
Conflict with management.

satisfied climatological interests (irrespective of logistical convenience), the National Science Foundation would not allow scientists to commission and supervise a recapitulation of the GISP-1 drill. Instead, NSF supported a group of CRREL veterans in an independent Polar Ice Coring Office, ordered PICO to design and build a new drill of unprecedented diameter for use on Greenland, and required the GISP-2 scientists to rely on PICO for the drill and logistics services. This policy sat so badly with the prestigious university scientists engaged in organizational efforts that they instituted an advisory committee to monitor the project:

None of us particularly liked the idea of a logistics group for major operation that wasn't entirely under the control of the operation. If people didn't perform . . . action to correct it had to be done through NSF. Essentially we always viewed the advisory committee as something that would meet in case there was trouble to add muscle to pushing the NSF around.

The scientists' suspicious attitude toward PICO, the isolation of living and working on a glacier, and the difficulties of solving technical problems in the field led to increasing contention between the scientists and PICO's field manager. "My manager," PICO's director concluded, "absolutely could not get along with this chief scientist, and of course there's only one winner in a case like this: it will be the chief scientist."

PICO replaced the field managers. The technical problems slowed but did not cripple the drilling. In retrospect, operations were successfully completed and the scientists benefited from PICO's participation. But this did not erase memories of resentment at having to rely on autonomous engineers for support:

> Had we been running the logistics, I probably would have had a heart attack for one thing, because we had enough work to do as it was, but had we run the logistics we . . . probably would have picked as good a design crew [for the drill], . . . [and] I personally feel that we could have saved a lot more money if the science had run the logistics. At one time I think PICO had 15 full-time people working on their staff in logistics. It's an immense number of people. It was larger than it had to be, and created a lot more friction.

By continuing to maintain an independent pool of managerial expertise for polar field science, NSF created grounds for conflicts between scientists and project managers.

The logistical difficulties of field work are not the only source of conflict. As figure 5.4 illustrates, *conflict between scientists and project management is associated with bureaucratic organization.* When participating scientists form a hierarchy with a designated leader, when a collaboration's leaders make important decisions with little collaboration-wide discussion, or when the project utilizes formal external reviews, conflicts between scientists and project management are more likely.[24]

AMPTE and IUE are two cases in which the formative phase generated scientific leaders who felt they embodied collaboration-wide scientific interests. As such, they were obligated and entitled to do battle with project management in the name of the science when they perceived central interests were at stake. The Active Magnetospheric Particle Tracer Experiment (AMPTE) was a space-science collaboration in which the participating scientists united behind scientific leadership and pressed for the development of capabilities (like near-real-time data processing), that the project manager believed unwise to pursue. Like AMPTE, the International Ultraviolet Explorer (IUE) had scientific leaders who outranked other participating scientists. NASA managers once threatened to shelve the project when one group of scientists appeared unlikely to meet the project's schedule for delivering a space-worthy camera with a mechanism for converting ultraviolet light to optical wavelengths.

In both cases, the scientists prevailed in part because their leaders could use external advisors to bring pressure to bear on project management. In theory, project management could have agreed to comply with the scientific leadership and avoided conflicts. That did not occur, nor is it likely when project managers are independent of scientists. But in these cases the two scientific leaders involved had no qualms about claiming that the scientific objectives depended on their specific requests. This contrasts with projects such as Voyager, Giotto, or the International Sun Earth Explorer, where the Project Scientist felt obliged to secure a consensus among the Principal Investigators before challenging the Project Manager—a consensus that could easily have failed because of the independence and conflicting interests among the Principal Investigators. Scientific leaders in hierarchical collaborations lead by challenging the authority of project management in high-level meetings from which the other participants are excluded. Success, for these collaborations, did not depend on avoiding conflict but on resolving conflicts productively.

By enlisting external boards for design reviews and approvals, they signal their desire to guard against technical failures. This typically occurs when accountability is a high priority either because collaborators are utilizing their organizations' resources or because their visibility is such that they will alarm their patrons if they fail. These conditions invite participants to argue strongly for certain preferences: few want to be held accountable for actions they feel are not in the best interest of a project. The external panels become adjudicators for the arguments about how much technical risk is reasonable to assume in order to achieve scientific goals.

Conflict between Scientists and Engineers

A third major structural potential for conflict exists between two major occupational statuses: scientist and engineer. Figure 5.5 shows that *conflict between scientists and engineers often emerges in tightly interdependent collaborations.* Where autonomy in instrumentation was low, at least half of these collaborations experienced scientist-engineer conflicts. This is in stark contrast to projects with medium or high autonomy in instrumentation, where in just less than 10 percent of the cases were there serious disagreements and tension between scientists and engineers. Similarly, where autonomy in the analysis of data was low, two-thirds of these projects experienced such conflict.[25] A variety of factors lead to these kinds of interdependence. In some collaborations, members must take into account the development of each other's instrumentation as they design their own. In others, members need one another's input in order to feel confident working with data taken by instrumentation designed and built by others. In such interdependent collaborations, scientists and engineers who find themselves in disagreement about instrumentation may appeal

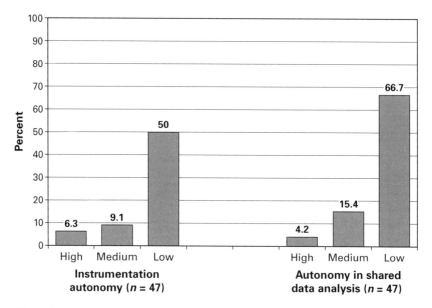

Figure 5.5
Conflict between scientists and engineers.

to the collaboration as a whole to settle their differences. Participants intervene to prevent disputants from losing sight of project goals. On the other hand, where collaborators are more autonomous, even if there is conflict between the scientists and engineers working on instrumentation, the conflict does not seem as significant.

The members of SLAC PEP 4-9, a technologically ambitious particle-physics collaboration, knew they would have to combine data from multiple detector components to produce meaningful results. They also knew they would have to elicit each other's input to be confident in their interpretations. As was typical of colliding-beam experiments in which detector components were concentrically nested around the point where the beams collided, the collaborators recognized their need to be cognizant of each other's technical and systems engineering constraints in the design of their individual components. The collaboration began without much conscious consideration of its organization, as though it could operate in the usual non-bureaucratic manner of particle-physics projects. The formal titles and the prestige they carried followed from the recognition of individual creativity. "Since the basic idea was X's," one participant recalled, "it was very natural that he be the spokesman."

However, the centerpiece of this project was the time projection chamber (TPC), a highly innovative component that added a new measurement to the capabilities of chambers for tracking charged particles. As the TPC's builders encountered technical problem after technical problem in scaling up their prototype, the leaders realized that an organizational hierarchy was needed to manage the hardware. More important, insight into the TPC's inner workings was not the prime criterion for position at the top of the hierarchy:

> We went from this table top thing that one person could carry to something that weighs a thousand tons in its entirety, and one of those extrapolations was the drift length. We discovered that the way we had made the structures that established the electric field had a flaw [which distorted the drift path of the electrons].... I personally worked for about a year on that problem trying to figure it out.... I remember trying to fix the electric field by meticulously and laboriously placing copper tape over the fiberglass structures on the inside.... Reality really set in with a vengeance after a while.

And the reality, as they came to see, was that the collaboration could not function with a titular leader who wanted to spend his time developing his prototype into a full-scale component:

> I was spokesman during the construction period. I thought I would step down at some time. I wound up stepping down earlier, because of exhaustion from trying to cope with all the things that were going on. I went and worked on the TPC itself. I have to admit I'm not the world's greatest manager. It just isn't my strength.

The realization that the spokesperson had to manage the collaboration led to the appointment of a physicist comfortable with empowering engineers to scrutinize physicists' plans. The participating physicists working on components that were incremental improvements on earlier designs did not like administration by engineers, but it was preferable to investing time in what could otherwise have been declared an outright failure.[26]

The view that conflict is largely a function of interdependence is at odds with the traditional—and, we believe, simplistic—view that such issues originate in differences in orientation toward technology that arise from the professional interests and values of two groups. True, such issues are charged with meaning, in view of the dependence of scientists on engineers for the technology that makes or breaks their research effort and the frequent lack of credit given for the instrumentation that results in discovery. We found that, although the interpretation of timetables and specifications[27] usually forms the framework for consensus or conflict between scientists and engineers, there is so much flexibility in the construal of professional "values and interests" that it is difficult to predict which particular view will be held and advocated by either group at a given time. The following perceptive comment by one scientist is worthy of any social constructivist:

> The thing that was interesting to me, that we hadn't thought of before and I wish we had, was acceptance of things. When does the collaboration accept the telescope as being . . . delivered. . . . When does it accept the spectrograph? There are certain specifications that the Scientific Advisory Committee had put on, say, a spectrograph or image or something. Does it really need it? . . . That was the source of disagreement[:] whether [the engineers] really delivered what they said they would deliver.

Experimental physicists are often viewed as interested in innovative instrumentation, taking pride in ceaselessly finding means to improve the capabilities of instrumentation. Engineers are often seen as interested in building instrumentation that meets specifications, taking pride in delivering the hardware on time and within budget. That was certainly how the astronomers working on the Keck Observatory saw the situation:

> But the astronomers and the design person . . . want it as good as can be. . . . And the guy who builds it wants it to be built within budget and on schedule. And his goal is to get it done; the other guy's goal is to get it perfect. And I'm not saying that they don't realize each other's problems, but . . . "We want it now," "We want it perfect," . . . and the other one is "We want to build it, make sure we have enough resources as quickly and cheaply as possible, so if we get stuck we have money to do it."

What we found, however, is that the reverse can also be true. Engineers sometimes view innovation as essential to achieving their objectives and interests. In these cases, they take pride in building the instrumentation as close to the ideal as possible, while the scientists, eager to start collecting data (or to obtain a secure source of data), want straightforward, frugal, no-frills instrumentation at low cost so that policy makers will not balk at funding the next project:

> On the one hand, the scientists want something which the engineers feel that they do not have the resources or schedule to deliver. And on the other hand, the engineers, and especially this group of engineers because they're really good, taking infinite pride in their work, want to take long enough to do the job right, and the scientist is there screaming "No, give me the goddamn telescope."

The managerial dimension of performance is only the *framework* for consensus between scientists and engineers. Acceptance of that framework is not the province of any one group: engineers may be either more or less averse than scientists to regulation by time and money, and are equally interested in success as they come to define it for a particular project.

The difference between traditional forms of small-group collaboration and multi-organizational collaborations is illuminated by the divergent ways in which conflict is managed. Particle physics throughout has been a model for the combination of intense teamwork, high complexity, and large scale that involves fluid recombination of participants. As a contrasting case of small-scale but extremely intense collaborations, consider string quartets. The performance interdependencies of string quartets are instantaneous and comprehensive, but, like other task-oriented work groups, string quartets integrate the contributions of individuals toward a common objective and often experience inter-group conflicts. What means are available for dealing with conflicts? Keith Murnighan and Donald Conlon (1991) identified a set of common strategies.[28] Groups abandon arguments and confrontations until later (or forget about them entirely). When differences do not disappear, they alternate between solutions, choosing one now and another one the next time conflict occurs. They defer to the leader (first violin). They define regions of silence, about which they will not talk. They emphasize the shared goal and not the differences in preferred means. Rather than "concede," they use one of the other strategies, and they define conflict as a positive value. All these strategies are used in scientific collaborations too, particularly the last two. The best ways to construct instrumentation, develop a technique, and collect data are viewed as emerging from the clash of perspectives. Without alternative opinions, how will the best way forward emerge? This is why scientific collaborations typically accentuate the shared objectives. As was noted above, collaborators often urge contenders to see their differences as irrelevant to project goals. Their common commitment *produces* differences (alternatives) that are, in essence, the means to these ends. "Concessions" need not be an issue when the chosen strategy is a product of many competing aims—costs, timings, and compatibilities must also be considered. Conflict, the dispute over technological means, is the path of knowledge.

Such means, on the other hand, do not suffice for managing organizational frictions and contentions that arise from the division of labor, from disputes over resources, communication, centralization, and hierarchy. The potentials of bureaucracy must be employed to this end. As we have seen, the organizations that collaborate exchange personnel, hire each other's students, redefine tasks, and sometimes minimize interaction between disputants. When subcontractors do not perform, they may be fired. Personnel who do not give their best effort may be replaced. Disputes may be anticipated before they actually occur. While technological conflict is subject to management through the traditional and argumentative means of self-contained groups, organizational conflict is a product of combining independent units and requires new methods and palliatives.

Performance

In the scientific collaborations we studied, trust was not associated with performance. The reason for trust's perceived importance is the association with conflict. What factors, then, contribute to the performance of scientific collaborations?[29] There is no more difficult question to answer, fundamentally owing to the wide variety of interpretations and criteria that may be brought to bear. Performance in terms of what? In our interviews throughout the 1990s, we recorded a wide variety of expressed views: carrying off an extremely complicated experiment, staying under budget while building a new instrument, cooperation of former competitors, sticking to a strict timetable through long hours. Moreover, scientific research is notoriously unpredictable. While spectacular new discoveries are often unplanned but unequivocally a hallmark of performance, moderate success or failure—characteristic of most scientific work—is far more difficult to evaluate.[30]

What seems certain is that scientific performance operates as a fluid and contingent concept. "Performance," together with "success," "effectiveness," "efficiency," and a variety of other notions, is one of an overlapping family of concepts whose similarity lies in their relationships within a class of explicitly evaluative dimensions of organizational life (Shenhav et al. 1994). In the following, we focus on its perceptual aspect. Discourse about "quality" is a fundamental aspect of all social worlds (Shrum 1996). In science and technology, concepts of "goodness" are employed pervasively both by analysts and participants. They are embedded in the discourse of social organization as an evaluative language applied to the processes and consequences of scientific work.

Judgments reflect an important aspect of the way participants relate to scientific knowledge and work. The choice of terms is not as important as our assessment of the varieties of meaning participants express in evaluative discourse. Such an assessment precedes a determination of the relationship between evaluation and other dimensions. No single, underlying concept unites these various meanings. No hidden factor underlies the concept of performance.³¹ Yet project managers and participants all evaluate scientific collaborations and tell stories about success or failure, about whether their project was better or worse than others.³² Moreover, they are keenly aware of the opinions of others—both scientists and funding agents—sometimes indicating their own judgments are at variance with those of others.

The most common types of judgment express two broad dimensions of performance in scientific collaborations, each of which consists of a pair of elements. The *managerial* dimension relates to the resource issues of time and money—to what degree did collaborations finish projects on time and within budget? Our informants were very much aware that these issues are subject to alternative constructions: What constitutes "on schedule"?³³ How much of an overrun is really "over budget"? Still, a project that lags far behind its original timetable and budget was generally described as by its leaders "over budget." Whether they considered these matters pivotal was something else altogether. A collaboration that spent far more money than originally intended, and finished well after its target date, might still be considered a success in terms of science.

The *success* dimension entails a distinction between the perceptions of participants and their perceptions of others who were not involved in the collaboration. Our starting point was the question of how "successful" participants considered their project *relative to others they had experienced*. We view it as "internally defined success," since no other actors are taken into account other than the participants themselves. A second element was how the collaboration appeared to outsiders—we refer to this as "external success."³⁴ We asked informants how successful their *peers* considered the project. Invariably, informants produced thoughtful responses that initiated a broader discussion of why they considered it a success, in terms of what factors, to what extent others in the field considered it successful, and what they thought accounted for the difference, if anything. The significance of the questions stemmed partly from the interpretive latitude they provided the informants. There is no doubt that the criteria informants used to judge the success of their collaborations varied—but the overall judgment of the success of the project varied little by informants within the same collaboration.³⁵

Time and Money

Two primary factors are associated with delays in completion: the presence of an external advisory committee and autonomy in the analysis of

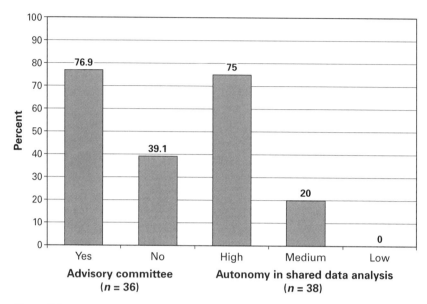

Figure 5.6
Delay in completion.

shared data. As figure 5.6 shows, more than 75 percent of all collaborations with an advisory committee *outside* the project itself experienced delays, versus 39 percent of those that did not utilize such a body. The association of delays with autonomy is even larger. The analytical phase, which occurs after data have been collected, requires more time for the coordination and integration of results when groups work independently. None of the collaborations with low autonomy reported delays. Of those with a moderate level of autonomy, one-fifth reported such delays. But in collaborations where teams possessed a high level of autonomy in the analysis of shared data, three-fourths finished later than their target date.

The second Greenland Ice Sheet Project (GISP-2), used above to illustrate conflicts between scientists and project management, exemplifies this combination of characteristics. External advisory committees do not themselves cause delays.[36] Rather, collaborations form an external advisory committee when they seem exceptionally important in terms of their scientific significance and risky in their technical requirements or ambitions. This state of affairs may well be undesirable from a strictly managerial standpoint—it means that powerful scientists are highly motivated to keep the project going in the face of difficulties of all sorts.[37] Projects such as GISP-2, carried out under treacherous conditions, are likely to encounter difficulties that seem worth solving outright. This is so even at the cost of delays *rather* than simply "adjusting" to these difficulties and compromising project goals.

As the project moved forward, many individuals worried about the funding agency's decision to appoint an independent contractor for both general project logistics and development of a new ice drill. The contractor would not be supervised by the scientists. For American scientists, who had not participated heavily in the climatological analyses of the GISP-1 ice core, GISP-2 was their opportunity to achieve state-of-the-art competence in dealing with what could be the preeminent source of paleoclimatological data. The GISP-2 advisory committee was formed "just in case." Important scientists in the collaboration distrusted the funding agency because of the way it had, in their view, mishandled the project's formation. They worried about the contractor's capabilities because it had not been the main builder of the GISP1 drill and had recently lost staff. If the agency or the contractor responded to project difficulties by doing something ill-advised, the advisory committee would be there to lead the fight.

In the field, numerous technical and personnel problems emerged. The drilling required an additional season, but the collaboration sought solutions and the advisory committee never became active:

We had been concerned . . . that maybe they [the logistics group] couldn't pull it off. They had to build the buildings, build the drill, run the drill, and do all that kind of stuff. It has worked out. It hasn't been perfect, but it has worked out far better than anybody expected.

In the event, the collaboration took longer than planned and longer than the European ice-drilling project.[38] The existence of the committee was a response to the project's perceived importance. Delays are due to specific obstacles, but projects that have such committees are more likely to view completion as overwhelmingly important. Program managers have difficulty in pulling the plug when delays occur.

While external advisory committees are a structural indicator of perceived significance, the ways that data are managed are directly significant for the likelihood a project will finish on time. Where there is greater autonomy in the analysis of shared data, timeliness is a difficult achievement:

The data was all to be put on the FTP Server [maintained by the GISP-2 Science Management Office] in as timely a fashion as possible. People interpret that differently, depending on the person. Any GISP-2 PI can call up any time of the night or day, access a data set without permission from anybody. If they wanted to use that data set for publication, or for presentation, then they had to discuss it with the person who generated that data set and make arrangements with them. . . . That gave people a lot of possibilities for looking at data sets that were not necessarily generated by their lab.

The collaboration as a whole was powerless to set deadlines for producing scientific works. Participants determined when their data were ready for collaboration-wide use, knowing that once on the server, all would have access. At a time when the drilling was going slowly and the participants might have to wait for ice samples to analyze, one scientist attempted to mobilize the rest to draft a collaboration-wide paper on the basis of ice samples already retrieved. The effort proved impossible and was never repeated:

> Everybody agreed to it. We collected the data sets. In fact we used smooth versions of the data sets so that people could still publish the original data sets. . . . And I tried to get input from people. . . . I eventually got comments from all but two of the people who were willing to contribute data. . . . They finally didn't even send their data except to the FTP Server. . . . I wanted them to write up the science component of their work. They wouldn't do it. . . . I guess I could have thrown out those one or two data sets, but I had already put six months into writing it. Then somebody didn't like the approach and rewrote the whole thing, and we finally both decided that so much time had gone by that the whole value of coming out with this cartoon-like approach, just showing some general ideas, was no longer valid.

The participants who did not contribute to this initiative seemed unjustifiably obstructionist to the leader of this effort, but it is likely they had reasons they preferred not to make known within the collaboration. Perhaps they were having difficulties with instrumentation they were using in their home laboratories to analyze ice samples. Or perhaps they were unable at the time to give GISP-2 their best effort because of other demands on their time and creativity. The pre-conditions for satisfying the timeliness criterion do not favor independent analysis of shared data. When individual participants can decide when their data are ready to be shared within the collaboration and disseminated more broadly, the collaboration as a whole cannot establish a schedule for producing scientific results.

Time is not money. The factors associated with cost overruns in scientific collaborations are not the same as those associated with delays. Figure 5.7 shows that *collaborations that rely heavily on subcontracting to acquire instrumentation, that find their initial results problematic or make midcourse changes in instrumentation are likely to exceed their budgets.* The under-

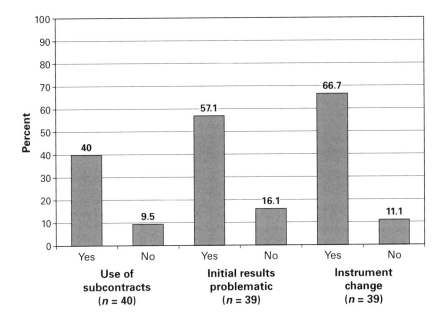

Figure 5.7
Budgetary overrun.

lying reason is that these collaborations cannot control their destiny through the straightforward deployment of resources and exercise of member competencies. These projects were less likely to be able to buy off-the-shelf equipment or make their own instrumentation. They could not guarantee the feasibility of their ambitions. They could not specify the source of unexpected or dismaying results. Such circumstances imposed the need to try mid-course corrections. The standard response to these conditions was to hope that a professional project manager could cope with the uncertainty. Sometimes it worked. In one case the participating scientists adhered to a division of labor that left the manager unencumbered, and the manager succeeded in keeping the project on budget:

> What helped a lot in the case of our project was the idea that we couldn't find the money. That is, we had a hard enough struggle finding the money we found. So it was vitally important in this project that the costs not overrun, because we didn't know where we'd find the additional money. The astronomers were willing to defer to the project manager because they felt that was important in terms of keeping the project under control. . . . You need a project manager who is separate from the scientists whose job is to keep the project under control. . . . When the scientists run the project, there is a built in conflict of interest that they always want to make it better, and the costs aren't an overriding consideration.

More commonly, however, project managers do not meet collaborative objectives while remaining within budget. Even when participating scientists are not inundating them with suggested improvements in designs or complaints about priorities, project managers cannot control what is beyond control, as a space scientist acknowledged when groping for reasons his instrument stopped returning data earlier than expected:

> When the [hardware] system fails, what do you do? You can say you should have had higher Q[uality] A[ssurance], but ... somebody said "Shit happens." And it does.

To overcome these possibilities, some collaborations spend more money. The scientific success of the International Ultraviolet Explorer, used to observe stellar objects from geosynchronous orbit for much longer than its projected life—rested on dubious accounting. Like many space-science projects, its American champions sought funding under NASA's "Explorer Program"—an ongoing line item in NASA's budget for relatively inexpensive science projects—to avoid the high-level politics of obtaining a new appropriation. But IUE's proposed detector system called for integrating three components that were individually unprecedented or infrequently used in satellites: an ultraviolet telescope with spectrographs that employed a kind of grating that had not been previously deemed suitable for space, an adaptation of an electronic tube (of a kind that corporations had just begun producing for the military) to convert ultraviolet to visible light, and sensitive television cameras with large enough apertures to take in the full spectra without scanning. Subcontracts were let for all three. The collaboration spent heavily as contractors for the converter tubes and the cameras discovered technical problems in making these components for use in space. While the manufacture of the spectrograph itself did not cause fiscal bleeding, the gratings themselves were so challenging that the contractor was funded to build three different focusing gratings for each spectrograph.

Lax accounting rules provided the money for coping with problems the collaboration courted by committing to integrating novel components that only subcontractors could produce. The American scientists who participated in research and development of IUE instrumentation were NASA employees. NASA engineers assembled the spacecraft and integrated its subsystems (including the scientific instrumentation). Technically, IUE stayed within the Explorer budget of $30 million to build the American portion of the satellite and operate it for a year—"but the wild card there was that civil service manpower was not explicitly accounted for. In a post facto accounting, the total NASA cost, when you add in the dollar equivalent of the manpower, probably was between $100 [million and] $150 million." Had university scientists overseen instrumentation and an aerospace company been retained to assemble and integrate the spacecraft, IUE would have caused a threefold-to-fivefold overrun on the Explorer budget line. Even if the collaboration had survived the furor it would presumably have inspired, subsequent evaluations of its success would have been affected.

Success

The second major dimension of performance is based on judgments of the relative success of the collaboration. Figure 5.8 shows that *participant views of success are shaped by the context of resource acquisition and the data agreements of teams*. What is significant about the former is that these factors concern the *formative* rather than the *concluding* stage of collaboration. The concept of "resource uncertainty" is used by organizational theorists as a way of describing a variety of external or contextual conditions that affect organizational structure and behavior (Pfeffer and Salancik 1978). In the context of our early analysis of high-energy physics, we did not view resource uncertainty as important, because the collaborations in our sample tapped standard sources of funding.

As the sample was expanded to include other fields, it became evident that many projects did not have a smooth route to the acquisition of resources. There was no clear funding program for submitting their work proposals and securing the funds required to execute the project. A reliable resource base for collaboration implies a funding agency with an established program into which the project fits comfortably, or a national research facility with the capabilities to support the kinds of investigations the collaboration proposed to undertake. "Routine" does not mean easy or certain by any means. Competition for routine resources can be brutal. Within an established funding program significant uncertainties generally surround *whether* a project will be funded. However, collaborations that justified themselves by claiming to be superior within a recognized research niche are distinct from collaborations that went hunting for resources with arguments that their ideas would generate a novel or under-appreciated research activity.

Figure 5.8 indicates the importance of this funding context in conditioning participants' view of success. Collaborations that begin under conditions of resource uncertainty are twice as likely to be judged as

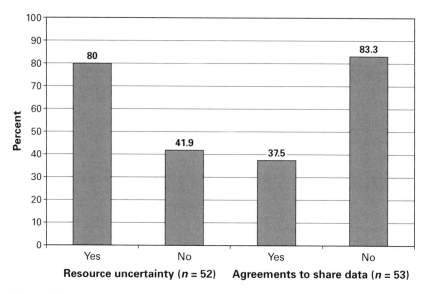

Figure 5.8
Percentage of participants describing a collaboration as "very successful.

successful by their participants than collaborations with relatively standard origins. When a collaboration was difficult to form—that is, when the participants had to overcome resistance to the very idea of a collaboration, rather than simply establish their particular claim on existing resources—then participants were much more likely to view their effort as successful. Relative to other projects in which they had participated, informants indicated that collaborations that were *uncertain or encumbered at the outset provided more positive outcomes*. This complex relationship has several sources.

Established funding programs generate similar classes of projects owing to (1) the existence of routinized budgets, (2) a range of expected funding amounts, (3) continual monitoring of these programs by the pool of past and potential recipients, and (4) the public nature of their products. When a collaboration is straightforward (i.e., has low resource uncertainty), its members have relatively high expectations. Such a project should exceed the accomplishments of other collaborations or their own smaller research projects. In particle physics, nearly all experimental research is done in collaborations. American experiments were funded by dedicated programs at the Department of Energy or the National Science Foundation. Our informants tended to compare the results they achieved to those of other collaborations. When their experiments did not produce impressive discoveries, they tended to downplay their accomplishments, even when they recognized that the causes of pedestrian results were completely beyond their control.[39]

When collaborations overcame obstacles to acquire the resources for research—that is, other than demonstrating their excellence in comparison to other similar projects—they were more likely to view themselves as highly successful. Three of the four collaborations that built ground-based astronomical observatories fit this description (Keck, HET, and BIMA). These were all "entrepreneurial" collaborations based on ideas that their instigators pursued for years before finding a combination of donors and organizations that made implementation possible. In evaluating these collaborations, participants did not speculate about other astronomical observatories. There was no central source of funding inspiring an entire community to generate similar prospects. But the judgment of participants was overwhelmingly favorable, even though in two of these cases (HET and BIMA) the collaborations were unable to adhere to their schedules or budgets for producing instrumentation.[40] Their most salient feature was that the collaborations provided their members with the research capabilities they promised, facilities as valuable when delivered as scientists had hoped in the early stages of the project.

The importance of this context of origin, particularly in funding programs, is evident in another way. Collaborations that formed in response to *new* agency programs, competing against an unconventional set of proposals for funding, were also considered highly successful by participants. Here, the process generated advantages that participants would not have enjoyed if the collaboration had not formed. They often considered the collaboration successful even without any outstanding result:

[I] would say it's been very successful . . . in terms of interacting with other researchers, in terms of having a bed of ideas for carrying out a research program. It's been very good in terms of the facilities and the talent it lets us draw in. So by the fact that we're associated with the center, that brings certain visibility and lets people know that good work happens at these places, and as a result we get better students and better postdocs and better visitors coming to us.

In responses to new programs, continuation is tied intimately to success stories. A failure to continue can be judged equally valuable when the collaboration ended because of a merger. One program manager used new legislative authority to encourage business competitors to work together. He considered a short-lived collaboration highly successful:

We started seeing things that led us to start mashing programs together . . . because you can see these guys . . . fit together. And they normally wouldn't [collaborate], so we sort of strong-arm the situation. . . . And that has worked very well. . . . On the business side, we have led in mashing together complimentary things on three occasions. Shortly after we mashed the programs together the companies consumed each other and merged.

The line of research was continued by many of the same scientists though the collaboration ended. Its major organizational pillars were no longer independent.

Independence in data analysis, as well as an encumbered funding context, is important to perceived success. The observatory-building collaborations and the collaborations that formed in response to new funding programs were similar in limiting collaboration authority over data acquisition and analysis. In the observatory collaborations where member organizations received observing time, each organization set up a committee to allocate time among proposals. Scientists who won observing time collected data without any obligation to share them, either with each other or with outsiders—though the collaboration posed no obstacles to individual observers that did choose to share data. Such arrangements befitted the general-purpose character of the observatories and ensured that the observatory would be used to investigate a wide variety of objects for a wide variety of purposes. Thus, the willingness of eligible scientists to use the observatory, as much as any particular result, implied success to the officials of these collaborations.

The collaborations that formed in response to new funding initiatives employed a related structure: they pooled the data-acquisition traditions of their constituent teams. Teams shared their results and planned further experiments in collaboration-wide workshops and reviews. Yet they rarely worked with each other's data and never without each other's supervision. The willingness of the teams to adapt their data-acquisition strategies or plans to fit collaboration goals were indicators of success to these collaborations.

In contrast, collaborations in which participants committed themselves to sharing data usually had leaders intent on achieving specific results as well as general aims. They judged the collaboration's success partly on how well they were able to achieve the specific result. For example, the Active Magnetospheric Particle Tracer Experiment (AMPTE) released tracer ions outside the magnetosphere and attempted to detect them inside the magnetosphere in order to characterize the role of the solar wind in giving rise to the charged particles inside the magnetosphere. This project required participants to pool their data sets in order to achieve this objective. In addition to the active experiment, however, the spacecraft orbiting inside the magnetosphere made many *in situ* measurements of the naturally occurring particles, and the spacecraft orbiting outside the magnetosphere released ions into the solar wind to create an artificial comet that could be observed both from Earth and with instrumentation on the spacecraft. The last two yielded results that largely satisfied the participants. But when asked to evaluate the project the following scientist could not shake his disappointment in what the data forced him to conclude:

We made the first release of lithium [outside the magnetosphere]. We didn't see a thing [inside the magnetosphere]. . . . We made the second release. We didn't see a thing. We analyzed data a lot. We were able to establish the upper limits on how much of the solar wind is getting into the magnetosphere, and those upper limits were a lot less than what the theories had predicted. . . . So, if you look at the original premise, to sit in the solar wind and measure what percent you had seen, that was a total and complete failure. . . . Eventually we came to understand that we succeeded in testing the theory, and we found out the theory was wrong. . . . [But] if I had known that I would see nothing, I would probably never have proposed the experiment. . . . The intrinsic assumption here was that we were going to make a measurement. It was not an either or proposition. We were going to say, well is it 1 percent, 10 percent, or is it 10–4 percent?

A collaboration that searched for resources to build general-purpose instrumentation seems highly successful once it is able to bring the instrumentation into existence. But when a collaboration integrates data streams in the interest of demonstrating the existence of particular phenomena or processes, the phenomena or processes must actually be measured for scientists to view it as successful.

The final aspect of performance is based on the second way in which scientists commonly speak of success: "in the eyes of others." These generalized "others" always know less about the "real" project than the participants. They are an imagined audience, the community of colleagues interested in the results of the project. Our informants sometimes noted that scientists outside the collaboration viewed their projects as more successful than they did themselves. We began to ask systematically how they thought outsiders judged their collaborations, a kind of "external success." Scientists' own perceptions of success and their view of outside evaluations were similar in three-fourths of our sample, diverging in the remainder. There was no pattern to the differences. When informants believed outsiders had a divergent view of their collaboration's success, these differences were equally split between higher internal success and higher external success. Scientists were as likely to see their collaboration as getting more credit than it deserved as to see it getting less than it deserved:

> [If] you want to be brutally frank about it, I believe the collaboration received a lot more attention worldwide than it really merited. But in terms of garnering recognition for the collaboration we did very well indeed. . . . I personally would have preferred to see the collaboration end much more soon than it did. I mean, this has been a long, long haul, and successful though it may have been, it just wasn't worth all this time. It really wasn't.

Where there was a perceived *lack* of appreciation from outsiders, they attributed it to divergent criteria for success. Divergent criteria can be the product of differences in judgment about the topic addressed with a community's standard range of instrumentation, or the differences in commitment to demonstrating the efficacy of particular instrumentation. An example of the former is a collaboration that used instrumentation common to high-energy physics experiments to examine processes that were more customarily of interest to nuclear physicists:

> I think high-energy physicists don't find this—I mean, they say it's successful but it's not interesting. . . . The problems are not like the problems in high-energy physics which are really very hard and require . . . enormous effort for just, you know, basically one discovery of one particle. In the relativistic heavy ions, you do a lot of things that nobody has ever seen before, but people might not think they are so exciting.

An example of the latter came from a medical-physics project that tested alternative technologies for diagnosing cancers:

> If the project, for instance, said that X was not useful compared to Y all the X community out there would be unhappy with it. And then we would say "No, X is very helpful relative to Y." The entire Y community gets unhappy about it. So there is no such thing as a happy ending unless you find positive results all the time. In this trial, like any other, we came up with results that were quite controversial at the time. For example, we showed that neither X nor Y were useful in staging prostate cancer. Well, believe me, that was not the most popular finding you could come out with. So we were attacked. You know, when you get attacked, you get attacked on everything. You know, your methodology and how come you are using this now—the technology is still evolving. That's the main criticism that you hear, "Why did you do a trial because the methodology is evolving?" Maybe we should wait until the year 3000.

When informants felt their collaborations had been judged relatively unsuccessful because other work was being judged more promising or productive, they explained their collaborations' apparent failings as due to coincidences of timing. Perhaps their competitors simply possessed superior resources, or perhaps they themselves had underestimated the challenge of turning their line of research into an ongoing tradition. Timing was especially important to particle physicists. The diversity of particle-physics experiments is constrained by the parameters of particle accelerators. When the accelerator produces few phenomena for study, the first collaboration to mount an experiment has a structural advantage:

If I look from the outside and say OK, the reason [is] we were the second type of this type of experiment at X. We started late . . . so we kind of suffer from that fact. . . . When you build a new experiment and a new accelerator, it's irrelevant what you measure. You're the first. . . . I think there are some highlights, there are some good data [in our experiment]. But the bread and butter part was not that good because our detector . . . was not the first one [to make those measurements].

Resource levels became a critical point in a collaboration that aimed to produce computation and communication tools with broader-than-scientific applicability:

I hate to be the most critical, but I think it is not going to be successful. . . . It was a very good idea 5 years ago, but industry has just taken over the collaboratory field and everything that's been done by X on a research level is pretty much available now to purchase from vendors.

A medical-physics collaboration that pulled together physicists from university departments and national laboratories and physicians from medical schools was foundering in the eyes of a physicist because physicians were not sufficiently cognizant of the collaboration's success:

It appears . . . likely that the project will not advance to the next level of doing 200 or 400 patients in a year, because we have no resources. . . . You have to get the medical community now to say "OK, it's a technology now, we see what it can do." . . . And they're not saying that in great numbers. It's OK, we've done our job as physicists, but you'd like to see it move on.

That outsiders might view some collaborations as less successful than participants themselves is a finding of little interest by itself. However, external and internal success are associated with different factors, which leads us to believe there may be alternative processes of evaluation at work. Figure 5.9 shows the percentage of collaborations that view themselves as successful from the standpoint of outsiders for three structural dimensions. *Large collaborations, those with international participation, and those with hierarchical decision making are more likely to be viewed as externally successful.* None of these factors are associated with internal success.

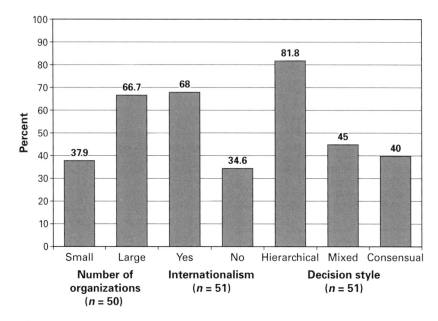

Figure 5.9
Percentage of non-participants describing a collaboration as "very successful."

Large collaborations and international collaborations are highly visible. The Giotto collaboration, for instance, was a pan-European effort to collect *in situ* data closer to Halley's Comet than all other such projects. On the night the spacecraft encountered the comet the project emerged as a public-relations spectacle. When the spacecraft and enough instruments survived the encounter long enough to collect data, outsiders were not going to care that some participants had jaundiced views of some of the design constraints that the project manager imposed from above on the scientists who developed instrumentation. Conversely, participants in smaller, national collaborations feel that their lack of visibility makes them seem less successful in the eyes of their peers no matter the satisfaction they found in participatory decision making.

At a more fundamental level, however, the relationship between decision-making style and success in the eyes of outsiders may be a consequence of the ability of the different organizations to generate results. The breadth of a collaboration's jurisdiction is usually much narrower when decision making is hierarchical and broader when decision making is participatory. The scientific productivity of hierarchically organized collaborations is usually a matter of enabling individuals or single-institution groups within the collaboration to pursue their research agendas autonomously. Outsiders may acknowledge the diverse results of participating scientists, who bask in their improved reputations. On the other hand, the scientific productivity of participatory collaboration is a matter of success in addressing those topics the participants agreed were worth investigating. Even if their instrumentation works as well as or better than expected, their results can still prove disappointing because the topics can't be addressed with the collaboration's instrumentation or because the topics don't open new lines of research or application.

Conclusion

Trust between individual scientists is decreasingly important in the scientific world, where multi-organizational collaborations are required. The central argument has been that the relations between structural components such as project teams are more significant than trust among individuals for the operation and success of large scientific projects. The central findings of the chapter are the following:

- Collaborators with prior ties have no greater overall trust than those without such relations.
- There is no significant relationship between trust and self-reported performance.
- Lower trust is associated with higher conflict between teams and higher conflict between participating scientists and project management.
- Larger collaborations and collaborations that provide more autonomy in analysis are less likely to experience conflicts between teams.[41]
- Collaborations in the field sciences and more bureaucratic collaborations are more likely to generate conflicts with project management.
- Greater interdependence in instrumentation and data analysis is associated with higher conflict between scientists and engineers.
- Delays tend to occur in collaborations with external advisory committees and those with more autonomy in the analysis of shared data.
- Collaborations that exceed their budgets have typically made mid-course changes in instrumentation and rely heavily on subcontracting.
- More successful projects, as viewed by participants, experienced uncertainty in the acquisition of project resources and avoided agreements to share data.

- Participants viewed larger, international projects as more successful in the eyes of their colleagues.

The chapter began by showing that collaborations based largely on pre-existing relationships did not have higher levels of trust. Large, complex projects sometimes require the cooperation of former competitors. Brokered collaborations are not uncommon. Such enterprises do not report less trust overall than those that involved friends and colleagues.[42] Why, then, is trust considered so important? Because trust is inversely associated with two of the three major axes of conflict: conflict between researchers and conflict with project management. Since conflict is oftentimes visible, personally significant, and distressing, the association between trust and conflict explains the continuing belief in the importance of trust. Yet the relationship between trust and conflict is complex, and certainly not unidirectional. Though projects with higher levels of trust exhibit lower levels of conflict, trust both affects and is affected by the flow of interactions and practices that can breed disputes or provide evidence for their existence.

Moreover, the degree of trust has no bearing on any aspect of performance assessed here. Projects that reported higher levels of trust did not report higher levels of performance for several reasons. There is a foundational aspect of trust that *is* important—trust that other scientists can contribute to a joint enterprise—but this foundational trust is so widespread that it does not discriminate between more or less successful projects. Trust fluctuates with changes in the goals and in the makeup of collaborative enterprises, which may have a connection with intermediate objectives (short-term outcomes) but not with the overall assessment of performance. Trust is inversely associated with conflict, which is higher in more bureaucratic collaborations.[43]

Projects with high trust are not seen as more successful by participants. What emerges as significant is that accounts of project success are conditioned more by the *formation* of collaboration than its conclusion. Where resources were acquired under routine conditions, a clever design that does not produce new phenomena or achieve a striking result is not viewed as successful. Under conditions of resource uncertainty, consumed at the outset with the tasks of generating support and finding funds, collaborators tend to see the *project process* as a success. In short, the institutionalization of collaboration makes Nature the culprit for poor results. Where human agency is involved, collaboration thrives.

The primary axes of conflict in research collaborations operate between structural groups and are associated with the *organization of their interactions*. To say that the interdependence of collaborations leads to conflict is not a tautology. All projects are not equally interdependent. Those collaborations that seek to manage the topics their members analyze are more likely to experience conflicts between teams. Collaborations that make decisions hierarchically are more likely to experience conflicts between scientists and management. Conflict between scientists and engineers typically occurs in collaborations with low autonomy in the analysis of shared data.

What emerges may be characterized as a group of findings related to the *interdependencies of cooperative scientific work*. Such a view is reinforced by the conditions associated with performance. Scientific performance is one of a family of "goodness" concepts that serve as flexible resources for those who make science policy, for project managers, and for economists. Each of these groups is professionally occupied with concepts that legitimize and aid in directing large expenditures for scientific research. In this chapter we addressed two main aspects of performance discourse: (1) the "managerial" dimension of time and money and (2) the "success" dimension. In both, certain forms of interdependence figure prominently and the complexity of scientific organization is revealed. Collaborations in which teams entered into agreements to share data were viewed as less successful by participants. Where autonomy in the analysis of these data was high, projects were typically long and drawn out—well beyond their original timetable. And hierarchically organized projects—those that generated managerial conflict—were viewed as higher in external success.

Other forms of interdependence are associated with consequences that are less surprising when seen as *consistent* with traditional views of individual scientists pursuing careers. Hence, collaborations that exert (or attempt to exert) control over the topics analyzed by their constituent teams generate conflict between these teams. This is so because control over data is a primary requirement for credit where massive effort has already gone into the acquisition of resources and instrumentation. Where decisions are made on a consensual basis there may be higher trust—but often that trust is simply inferred from the consensual basis of decision making. It is conceptually problematic and empirically inaccurate (with respect to our data, at least) to draw any further implications as to the role of trust in the performance of inter-organizational collaborations. Trust both is and is not important. It is important because projects with more trust have less conflict—and conflict is aversive. But it is not crucial, since they are no less successful, in the eyes of participants.

The reason for this "paradox of trust" is that bureaucratic organization segments scientific work, imposing a structure for interaction resembling work that is actually non-collaborative. Of course, interdependency inheres in inter-organizational collaborations. Simple agreements to exchange information or the discussion of mutual interests at a conference are collaborative in a trivial sense. But the kinds of large-scale projects we have analyzed here are difficult to form and fund, time consuming for their participants, and awkward to terminate prematurely. Decisions about the relations between major structural components—between scientific teams, between scientists and project managers, between scientists and engineers—remain negotiable but have ramifications throughout the life of a project. The most "collaborative" collaborations occur in the field of particle physics, where decisions are often made consensually, without formal bureaucracy. For this reason, the field is of enormous sociological interest. As the preceding chapters have shown, it is the exception and not the rule.

Conclusion

Modern societies depend on knowledge as human capital, as a component of and a constraint on government policy, and as a source of technological innovation. Since World War II, the financial requirements, risk, and visibility of many research endeavors have generated public controversy and political debate. Scientific knowledge is accorded the highest status among various forms of knowledge (Boehme 1997), and it is important to understand developments such as the increasing prominence of collaborations in as much detail—but also as much generality—as possible. Historians, sociologists, and anthropologists of science have demonstrated that understanding scientific and technological change is no longer simply a matter of scrutinizing individual scientists; it depends on understanding the structure and dynamics of the institutions in which knowledge is generated. Processes involving external support relationships, consensus formation, and organizational status are critical.

Today, the resources of single organizations often are unable to satisfy the desires of those who are professionally committed to the production of knowledge. The elemental forces of nature, the history of the cosmos, and global climate change are too complex and costly to be tackled by an individual organization. Likewise, understanding the structure of collaborations is crucial for extending our knowledge of science in the contemporary world. The historian Peter Galison distinguished between the "outer laboratory" (the macroenvironment that encloses the experimental physicist) and the "inner laboratory" (the microenvironment or immediate surroundings of the working scientist). The importance of this distinction lies in the claim that the interactions of microenvironments and macroenvironments are essential to understanding pre-World War II physical research (Galison 1997: 3). But it is just this distinction that lost force in the latter half of the twentieth century as researchers developed "meso-environments"—temporary alliances of independent laboratories with researchers often working at different sites—to pursue their ambitions. "It has been," Galison writes (ibid.: 5), "a long, irregular, and often broken road between a time when it was unthinkable that a physicist be anything *but* someone who built equipment, designed procedures, manipulated experiments, wrote up results, and analyzed them theoretically to a time when it would be a matter of near-universal consent that someone could count himself (or, more rarely, herself) as an experimenter while remaining in front of a computer screen a thousand miles from the instrument itself."

Not only has the research process undergone significant alteration; so have the requirements for assembling and producing the conditions for discovery and innovation. In an underappreciated volume published in 1971, Leonard Sayles and Margaret Chandler noted the complexity of the problems created by the development of new technologies that produce and require collaboration among organizations. Inherent tensions were noted between the unrestricted search for knowledge and the requisites of inter-organizational collaboration, between program needs and institutional needs, between focused disciplinary orientations and more generic interdisciplinary demands, between temporary and permanent organizational arrangements. In our terms these are negotiated features of social organization in collaboration. Our task has been to seek regularities in the arrangements that researchers have adopted for generating new knowledge about some of our era's most vexing and intriguing scientific issues. This concluding chapter returns to the themes of technology and bureaucracy, the dimensions most crucial for an analysis of the structure of collaboration. We consider the differences—more important, the lack of differences—between scientific fields and the question of the necessity of collaborations.

Technology and Bureaucracy

From the moment they are conceived, collaborations and their leaders must confront the twin imperatives of bureaucracy and technology. Collaborations cannot succeed without attention to organization and management, even when they are non-hierarchical and participatory. Collaborations cannot succeed without the necessary instrumentation, even if they seek to employ standard, "off-the-shelf" equipment. Mediating, harmonizing, or, in the happiest cases, creating a positive synergy between the two is the essential drama in the evolution of most collaborations.[1] In short, collaboration is inherently "technoscientific." To understand the dynamics of any particular collaboration, it is crucial to ask how the design and construction of instruments interact with the organization of the participants. Our general argument is that the structure of scientific collaborations must be viewed in terms of the instruments and practices utilized in the acquisition and manipulation of information, practices that either integrate or separate organizational elements, shaping their autonomy in producing knowledge.

Several factors support this line of reasoning. The technological "imperative" is frequently the focus in studies of cooperative arrangements in science. Both Zabusky (1995) and Knorr Cetina (1999) describe the ways technical objects have the ability to shape group structure and management, in space science and particle physics respectively. The sole objective of a collaborative venture may be the construction of a suite of equipment, as in the telescope-building projects of ground-based astronomy. Generally, though not always, this introduces a structural distinction between builders and users, with different timetables, orientations, and interests. Even where instrumentation per se is not the primary objective, the scarcity of particular instruments often make collaboration necessary for particular research objectives. Inter-organizational projects are socially organized and managed in ways that must take account of (1) the use of technological facilities to acquire data and (2) the interests of collaborative building blocks—individual scientists, teams, university departments, research institutes, government agencies, research laboratories, corporate R&D laboratories.

The scientists and engineers we interviewed over the years found technological determinism an agreeable concept. Not so with bureaucracy. They preferred to view social relations as products of technical needs and research ambitions. Certainly, participants in collaborations that required large scale or highly innovative instrumentation required confidence that their organization was appropriate to such daunting tasks. But an embedded prejudice casts doubt on the explanatory power of technological determinism. Who among us, in a taped interview with a stranger, would not prefer to explain professional relationships in a manner that skirts direct discussions of morality and power? Even with this structural inhibition, it was clear that political imperatives shaped collaborations whose activities seemed likely to affect the interests or standings of their member organizations. Management would have to solve, diffuse, sidestep, or tolerate conflicts among member organizations.

The tension embedded in scientific collaborations arises from two conditions. First, they are voluntary in character. Second, there are opportunity costs associated with collaborations: no one has to join and everyone could be doing something else. Creating a structure is a ticklish business, feasible only because these projects are temporary and because the data they may collect are so alluring. In a fundamental sense, the desire of the constituents to protect autonomy and to prevent any one unit from imposing its interests on the others is both a barrier and an incentive to imposing bureaucratic practices within collaborations. When participants do not see their interests as compatible with each other and the collaboration as a whole, they must balance the protection provided by bureaucratic structures against their distaste for bureaucratic procedures and their fear that the collaboration as a whole could become too powerful.

When a central authority such as NASA selects collaborators impersonally, when organizations with histories of competition or little familiarity choose to collaborate, members tend to create formal procedures and hierarchies to circumscribe the collaboration's jurisdiction and to ensure fairness and accountability in decision making. When collaborations are formed by organizations with cooperative histories or by organizations for whom collaborating is essential for research (e.g., university physics departments with particle-physics groups), members were more likely to avoid formality and hierarchy in favor of arrangements that mirrored past practice, assuming members would behave honorably and reasonably.

The existence of political imperatives, however, should not mask the importance of technological and scientific considerations in the shaping of collaborations. Any group—whether a temporary collaboration formed for a limited purpose or a longstanding organization with no intention of ever outliving its usefulness—that intends to buy large amounts of instrumentation is well advised to empower a central authority to oversee contracts and purchase orders. Any outfit that seeks to integrate a technological innovation with more conventional instrumentation or to confront exceptional logistical challenges should employ professional project management to deal with the difficulties it courts in budget and scheduling. Any project that seeks to operate multiple instruments in a coordinated fashion or to merge data streams must shun formalities that might lead participants to object to their influence on the management and use of data.

How well political imperatives and technological practices mesh have significant consequences for collaboration. Individual participants in projects were satisfied when they used bureaucratic structures to protect the autonomy of member organizations *and* could produce knowledge by giving each of their teams the autonomy to integrate their techniques with the resources provided by the collaboration. Individuals were satisfied when participatory management made for productive mergers of data streams *and* when collaboration activities did not affect their parent organizations' interests or their interests within their parent organizations. Mismatches between levels of bureaucracy and technological practices were at the root of much discontent for the collaborations in our sample. While no collaboration we studied disintegrated, in some cases the collaborators could have designed better arrangements to avoid creating conditions for conflicts. In others, they were irreducibly stuck with conflict-provoking conditions.

Particle physicists usually spoke glowingly of how their participatory, non-hierarchical management made collaboration-wide physics more important than the interests of member organizations. As a general characterization, there is no reason to dispute this image. The obvious question arises: Should collaborations in other disciplines seek to be more like particle physics? Our findings indicate that they should not and probably cannot, for three reasons.

First, particle-physics collaborations can manage themselves as quasi-Athenian democracies under circumstances that are relatively peculiar—the paucity of options for non-collaborative work, competition for the use of scarce accelerator laboratories, and the scientific imperative of combining findings from separately built detector components to reconstruct events all help to make participatory management both possible and desirable. Because graduate students in experimental particle physics earn their doctorates by working on collaborative experiments, particle physics has developed a self-sustaining culture of collaboration of a sort that is not available in other fields.

Second, this image of particle physics often involves a confusion of collaboration and cooperation, with the existence of the former taken as evidence of the latter. Evidence of conflicts involving much more than simple intellectual disagreement is not hard to find. When a junior faculty member (or, worse, multiple junior faculty members) considered a collaboration experiment the centerpiece of a quest for academic tenure, conflicts were generated in the attempt to acquire personal credit or boost the role of the junior faculty's organization. In such cases, the political imperative of creating a case for tenure within a member organization was mismatched with technological practice, which obliterated the importance of member organizations by merging data streams and subjecting all findings to collaboration-wide review.

Third, individual particle-physics collaborations occasionally departed from the particle-physics norm when they incorporated technological innovations and found that stronger project management was needed, or when they became so large that middle managers were needed to filter the work of members before receiving attention from the collaboration as a whole. Though both conditions created more hierarchy, the former was more difficult for particle physicists to accept, since it invested power in an "outsider." However, it brought these projects closer to the structure of space-science collaborations, in which a project manager appointed by NASA or by ESA has ultimate sign-off authority on the budget, the schedule, and the engineering of the spacecraft and its suite of scientific instruments. This structure worked for space-science collaborations when their technological practice was to integrate instruments individually with the spacecraft and to emphasize scientific topics on which individual data streams shed light. When space scientists have attempted to operate in "particle physics fashion" without the technological practices of particle physics, they have generated unproductive conflict with managing engineers.

Except for particle physics, the specialties in our sample did not display dominant styles of collaborating. Even space-science collaborations, which had a nominal structure imposed on them by NASA and ESA, succeeded in resisting or modifying that structure when it conflicted with their technological practices. Several geophysics collaborations in our sample better approximated the nominal space-science structure than most of the space-science collaborations themselves. When geophysicists wished to take data in remote, harsh, or dangerous locations, they resorted to a hierarchy in which a professional took charge of the logistics and member scientists designed instruments and practices to fit the context that the project manager established. To the extent that some members sought to combine data or samples to address a particular topic, they were challenged to overcome a bureaucracy oriented toward protecting the autonomy of individual science teams.

In other cases, geophysicists created collaborations that employed highly formal procedures rather than powerful project managers to ensure that the collaboration did not become an extension of any member organization. For example, with the defense and intelligence communities spending liberally for better ways to detect nuclear explosions and the energy industry providing resources to detect and extract oil, academic geophysicists in the United States found themselves envying the superior capabilities of their government and industrial counterparts. Unable to command similar resources through their home organizations and fearful of any multi-organizational arrangement that might give one organization an advantage in obtaining and using collaboratively acquired facilities, academic geophysicists have created large collaborations that used rules and contracts to ensure that representative committees of members addressed the problems of obtaining and managing resources. Such arrangements worked so long as the practices of researchers favored a firm separation between individual scientific interests and generic instrumentation that the collaboration provided. The arrangements bred discontent when the collaboration itself set overarching scientific policies to which members had to conform.

Collaborations in materials science, more than collaborations in any other field examined here, protected the autonomy of their member organizations because they included organizations from multiple sectors (academic, industrial, and governmental). Usually the most important part of this protection was a formal intellectual-property agreement that carved out areas in which participants could share information freely without violating their employers' interests. Hierarchies were noticeably present when collaborations provided instrumentation for autonomous scientific teams, and were noticeably absent when collaborations merged data streams in an effort to understand particular materials. Whether hierarchical or not, the difficulties collaborators encountered in initially negotiating intellectual-property agreements inhibited them from renegotiating the agreements. Materials-science collaborations tend to expire at the end of their initial funding.

Astronomy collaborations that built telescopes were equally adept at protecting their member organizations. In some cases, histories of intense competition in the development and use of telescopes made this a political imperative. But even in the absence of such history, telescope-building collaborations choose to operate its telescope as if it was owned by each member organization for a percentage of its observing time. Hence, these collaborations vested usage issues in the individual organizations, which formed "Time Allocation Committees" to decide who from within the organization would utilize shares of observing time. Individual scientists from multiple organizations could choose to propose joint observations, but they were not especially encouraged to do so. In contrast, astronomy collaborations that performed the very-long-baseline-interferometry observations that were needed to merge the data streams of multiple observatories to obtain scientifically meaningful results and could not obtain VLBI data except by collaborating. Like particle physicists who formed collaborations and applied for experimental time on accelerators, astronomers performing VLBI observations formed minimally bureaucratic collaborations to equalize the rights of participants in the use and interpretation of the data to which all organizational members contributed equally.

Much history and sociology of science and technology has called attention to the importance of disciplinary traditions and communities in the practice of research. On the basis of the information we gathered and analyzed, we are most impressed with the *ir*relevance of discipline to multi-organizational collaborations. Their organizational and technological practices are largely functions of conditions that exist in all the specialties covered here. Most important, collaborations from all disciplines display a fundamental condition: *the more bureaucratized a collaboration, the more circumscribed its jurisdiction.* Collaborations with hierarchies and formal rules do not manage the activities that directly build the research reputations of individuals and organizations. Collaborations that vet the claims of their members prior to dissemination tend not to have formal rules and statutory hierarchies.

Value and Necessity

By the 1990s, collaborations were so numerous and so visible that collaboration had come to be valued "for its own sake" (Duque et al. 2005). We do not share this view. Multi-organizational collaborations are necessary when the resources of single organizations are insufficient for specific goals. That collaborative work might somehow be preferable to other, presumably more individualistic ways of producing knowledge is based on the notion that synergies occur when participants work together, resulting in greater productivity or efficiencies. But that is to consider benefits without costs, as if there were no resources required to manage collaborations, no meetings and communications involved, and no time and energy devoted to teamwork and conflict management. Without clear objectives, we see no reason to commit to collaboration as a mode of work and no reason to promote collaboration as a value.

The expense and scarcity of certain instruments is the most prominent of the conditions that make inter-organizational collaboration necessary. For certain fields, there is no practical alternative to collaboration and there are few models of how to do science *except* in a collaborative context.[2] In contemporary experimental particle physics, the socialization process requires that students function within a collaboration. Even design and construction of a detector component or prototype—what we typically think of as individualistic bench science—must be performed with the goal of integration with other components that are designed and built in other organizations. The borrowing that involves replication of social structural arrangements is second nature, since there is, effectively, no way to proceed without sharing of facilities owing to the resources required to build and operate them. Technology is, in this sense, the source of collaborative necessity. But there are still questions pertaining to the "social construction of necessity." What keeps research organizations too small to command the resources for particle-physics experiments? When do participants view collaborations as required for scientific work? Are collaborations actually necessary when scarce, expensive, large-scale instrumentation is not involved? Future studies will profit from close attention to the ways in which collaborative *need* is defined and how it evolves.

If it were the case that large instruments themselves were the source of collaborative necessity, then surely ground-based astronomy, not particle physics, would be the paradigm of collaboration. Designing, building, and operating telescopes and their accompanying suites of detectors were large-scale, multi-disciplinary undertakings when astounding experiments on radioactivity were within the means of a Polish woman working in a French university. Yet there remains a common sentiment among astronomers that the instrument is not the science. It receives overt expression by those whose work is most dependent on the technology. One participant described astrophysics as small science on large facilities, with the important work still done by teams of two or three researchers. Another claimed explicitly that a collaboration to build a telescope had nothing to do with the scientific results at all!

In recent decades, academic astronomers have collaborated to create new observatories because their universities could not afford the technology (and technologists) they considered essential for meaningful research. But even when they organized collaborations to meet their technological needs, astronomers did not adopt particle-physics-style technological practices or govern themselves as Athenian democracies. The use of technology as an explanatory tool should not obscure this important distinction, and future students of collaboration should focus on the nature of the connection between collaborating and the generation of scientific results.

Collaborative "necessity" in the absence of developing or deploying instrumentation proved an elusive concept in our interviews and tended to focus on relationship building, shifts in orientation, and "competitive" collaboration. One positive consequence of collaboration often mentioned by our informants is the development of friendships. A group begins to work together, international partners come on board, camaraderie develops, and relationships develop that last many years. But such opportunities to create friendships will not do as a justification for the notion of collaborative necessity. Social relationships produced from organizational interaction evolve in different ways and friendship is only one possible outcome. It would be fine, indeed, if all collaborations were cooperative. But collaborators often discover they dislike each other, friends can become enemies when they want their collaboration to move in incompatible directions, and time spent cultivating people in other organizations is time not spent cultivating people elsewhere within one's own organization. The costs and benefits of social relationships are important, but are not crucial to the question of whether collaboration should be promoted, unless the cultivation of relationships between researchers in different organizations is viewed as more important than conducting research efficiently.[3]

Through our interviews with government and laboratory officials, it was clear that program managers sought to promote the production of knowledge through collaboration, following the principle that more interaction is better than less. But interaction is not the same as collaboration. It is entirely too easy to confuse the intrinsic value of cooperation (in our terms, taking the interests of others into account) with collaboration (social organization oriented toward some objective). One of the most frequent benefits of collaboration mentioned by interviewees is simply the exchange of information via meetings. Investigators might meet three or four times a year throughout their project, exchanging ideas, concept, and advice that are viewed as more valuable than information received at standard scientific meetings. Yet in descriptions of such meetings, justificatory comments emerged—for example, "While I can't say that as a result of this I have a specific product, I'm sure that the scientific benefit has been great." The benefit of communication may be significant, but need a formal collaboration be constituted for such communication to take place? Any focused research area may profit from regular meetings between investigators, and, by implication, from resources to ensure that regular meetings occur. It seems reasonable to call formal collaboration *necessary* only when more than classic scientific teamwork—e.g., trading ideas, comparing techniques, swapping samples, critiquing manuscripts—is in the offing. Some of our informants saw collaboration as old wine in new bottles, doing what they would be doing anyway. The joint endeavor brings a greater sense of cooperation, but participants could each have achieved as much individually without it. Benefits were still perceived and celebrated, if only for the enhanced mutual understanding of scientific work, but this owed more to meetings and interaction than interdependence. Still, these perceptions were unusual.

Another type of "necessity" claim pertains to shifts of direction, when outsiders seek to change the perspectives or the problem selection of individual scientists. In some projects, managers are responsible for channeling researchers into the collaboration's structure. These are the "mini-funding agencies" where interaction is required. The "necessity" arises here through the discovery of common interests. The objectives are not established in advance, except in a kind of general way—to become acquainted with each other's concerns and capabilities and see what would emerge. Individual researchers join the collaboration and receive collaboration funding by proposing research that collaboration administrators judge to be within the collaboration's jurisdiction and judge to be a contribution to the synergy of interests and techniques that the collaboration is supposed to foster. Individuals have some leeway—they may choose with whom to collaborate—but the collaboration is a requirement of the continuation of funds, and scientists are well aware of it. To use the old saw, it is like leading a horse to water—it is not clear that such collaborations had an impact beyond the exchange of information and ideas. Again, we do not dispute the value of that exchange, but information and ideas are routinely traded in scientific networks. This is a peculiar form of collaborative work, a raw testament to the positive valuation of collaboration more than interdependent activities toward a common objective.

Collaboration can also be a "necessity" in a political and competitive sense. In the absence of collaboration, the interests and objectives of a group of scientists would not rate in the competition for limited resources. A consortium is formed to make use of emergent or existing instrumentation. Members of the community well know that it is going to be used; someone will figure out a way to employ the equipment for new scientific purposes and they will get the credit for it. Capacity is not unlimited and other opportunities are sometimes scarce. This kind of necessity is perceived by research specialties that do not obviously and directly impinge on utilitarian interests but must compete for resources with specialties that do.

Because collaborations are not always necessary, it behooves us to ask whether they can ever be *un*desirable. Because collaborations entail the expenditure of resources and opportunity costs, we have argued that collaborations should not be viewed as ends in themselves; the time and energy of participants could well be invested elsewhere. Policy makers and research administrators no less than research scientists themselves should be careful not to allow the improvements in administrative efficiency made possible by electronic communication to blind them to the very real possibility that more valuable research could best be pursued through the non-collaborative efforts of individual organizations.

The question of negative impacts reminds us of the obstacles to initiating collaborations and anticipates the difficulties of ending them. Collaborations roil scientific communities when their formation or their duration violates scientists' sense of justice and equity. If membership has its advantages, non-membership has its disadvantages. Particle physicists, who always collaborate, know that opportunities to submit proposals for beamtime at accelerator laboratories regularly provide moments for outsiders to become insiders and vice-versa. NASA and ESA, which fund collaborations, usually centralize decisions in order to equalize opportunities for participation. But occasionally a collaboration forms by word of mouth in a specialty where collaboration is not the norm, and the result is a sense of grievance within a scientific community. Some of our informants reported a sense of envy or jealousy among other groups working in the area that collaborators were getting more than their share of resources. Those who move into a coordinating role, in particular, may use, and be seen as using, facilities that others consider collective property of a community.

Some collaborations, whether by choice or by contractual terms, archive their data for others to use. But when data are expensive to obtain there are generally arguments about whether those who did not collect the data are suited to analyze them. These participants feel there is a strong case for continuing to fund data analysis by collaboration members in order to maximize the scientific return on the investment in acquiring the data. When such funding is generous and enduring it is likely to engender a sense of grievance. When a single coordinating organization receives long-term funding, reduced resources for other research groups creates opposition from within the field itself. The art of program management in an era of enthusiasm for collaborations must be to avoid these sorts of situations and to deal forthrightly with such situations when they do occur.

"Necessity," then, is attributed to collaboration in a variety of ways. Though it is most often associated with large, complex technologies, to speak of the importance of technology is neither to say that collaborations are formed exclusively to build instruments, nor that technical properties of these instruments determine the structure of collaboration. Instead, the kinds of interdependencies configured by technological choices, often made early in the history of a collaboration, shape subsequent developments. Needs can be social (as in collaborations that stimulate working relationships among specialties that do not usually work together) or political (as in collaborations that gain access to resources that would seem unjustifiable if proposed by individual investigators). Collaborations may, but need not, involve high levels of cooperation, and can end in enmity as well as in friendship. Under extreme but not far-fetched circumstances, extended collaborations can be detrimental to the overall morale of scientific communities.

How Collaborations Work

For those who prefer to read the end of a book first, there is good news, for this end is also a beginning. We began this volume by arguing that the concept of "Big Science" has outlived its usefulness. Crescive levels of aggregation in modern research networks render it more important to examine and compare structural elements (collaborations, organizations, teams) or temporally defined occurrences (discoveries, measurements, results). Our unit of study was a project that involved a minimum of three organizations. Magnitude—our preferred term to describe both scale and temporality—can be viewed in several ways: the number of participants, teams, or organizations; the time it takes to acquire funding or to produce a publication.

What remains of value from the notion of "Big Science" elaborated by Price and Weinberg is that larger projects are more likely to be shaped in conjunction with features of the articulated policies of a funding agency (public or private) and the organizations that employ scientists. What is most interesting is that although larger collaborations utilize more elaborate, formalized structures, they do not employ hierarchical decision making about matters pertaining to the generation of scientific results. "Facts and findings" are treated in an egalitarian fashion when reviewed by an entire collaboration or assessed within smaller structural units, which disseminate their findings autonomously, without much regard for the collaboration as a whole. The organizational constraints of large collaborations do not make them unsociable knowledge factories that overwhelm or undermine skepticism and creativity. While technology is implicated in almost every aspect of collaboration, participants view the process of result-making as even-handed or autonomous. Adjustments and adaptations render the size of collaborations much less significant than we expected at the outset. For resource decisions, including matters of efficiency and budget, size effects are not straightforward. Oligarchies emerge in collaborations with larger numbers of participants to make resource decisions. Corporate boards of non-participants become involved as the time frame increases. Increasing the number of participants affects the procedures for managerial decision making. It does not much affect decisions about research strategy or the evaluation of results.

The formation of collaborations involves the creation of a context that members expect to serve them better than their alternatives. For some, alternatives may be as simple as small projects and fewer meetings. For others, the alternative may virtually be non-participation in research. Science and technology scholars generally—and historians especially— seek to identify the contexts in which knowledge is created in order to characterize particular times and places. Codifying the formative stories of multi-organizational collaborations thus promises to be an excellent starting point for identifying essential features of science and technology in the twenty-first century. For managers who want their collaborations to proceed smoothly and terminate gracefully, an analysis of collaborative formations leads quickly to the question: Does the way a collaboration forms constrain its evolution? Are there regularities that should guide the instigators as they confront how best to make their ambitions and expectations a foundation for recruiting participants, developing a system of collaboration governance, and carrying out the scientific work?

Our results support the broad hypothesis that the formation of social organization has an important impact on operations and outcomes—an effect that occurs through the contingencies of historical process rather than any simple structural clause in their constitution. While it might seem contrary to common sense, the beginning of a project may have more to do with its success than its end. Collaborations that struggle to acquire resources, with higher levels of uncertainty in their formative stages, are viewed as more successful. Resource uncertainty creates a context in which "making it over the hump" is decisive for later developments. Submitting a grant proposal in a routine competition for funds involves uncertainty, to be sure, but that uncertainty is routine. Scrambling, strategizing, meeting, and marshalling support generates a sense of success *early* in the life of a project that carries through its later stages. The group of collaborations studied here consists solely of those that were funded, but it seems unlikely that a broader sample, including those that did not make it, would refute this moral.

Patterns of formation are more readily characterized by complexity and sectoral composition than scientific field. The acquisition of resources for some involves nothing more than a successful grant proposal, while others take years of beating the bushes, self-funding, state start-up funds, or private philanthropy. Collaborations do not simply consist of groups of friends or acquaintances. Many include participants who do not know each other beforehand. Because of the financial resources required for some projects, program managers may have an important role in their formation. Some projects do not have any single entrepreneur or initiator; they may involve a phase of selecting participants through an impersonal competition. Some begin with struggles and conflict over roles by institutions that will subsequently work together. The breadth of participation is especially significant when collaborations form without any dominant sector but include academic, private, and state agencies. Unconventional combinations of organizations—a "triple helix"—will involve negotiations among different traditions and constraints (Etzkowitz and Leydesdorff 2000).

The range of motivations is equally large. Some initiators want to learn new skills and push the envelope of novel research technologies. Some seek to rekindle their careers by hitching their wagon to a new set of horses. Some participants respond to pressure from their employing organizations. Some seek the kind of consistent collaborative involvement that may be generated from networking with colleagues, developing a reputation as a competent and valued member of a large team. Because of the impossibility of bench science and the importance of future participation in the ongoing flow of collaborations, particle physics in particular is characterized by such motivations (Knorr Cetina 1999). Scientists in this field, their employing organizations, and the accelerator laboratories that serve as episodic hosts, understand that collaboration is the essence of the discipline and socialize neophytes appropriately.

From this diversity of agents, organizations, and motivations five main processes of formation were identified. The first was characterized as conventional, since it involves academics who previously worked together, undertaking a project that is too large and complex for a single team of scientists. An initiator may communicate the need to "wake up and smell the coffee." A second process is similar in its relative homogeneity and conventional acquisition of funds, but these projects coalesce around participants with little history of working together. Often the context is one of significant struggle ("Damn the torpedoes!"); these projects are less likely to have a single initiator. A third variety emerges from a group of scientists with pre-existing relationships but no obvious source of support. They face pressures from their employers, sometimes opposition from fellow employees, but manage to "get out of the rut" with a project that often creates a new facility or a novel alignment of scientific interests. The fourth process, like the second, involves participants without pre-existing relations, but these collaborations are assembled through impersonal selection processes and external brokerage. Finally, in the "collaborators wanted" model, scientists with a shared vision bring together others with valued competencies, in a conventional resource framework. Participants need not have worked together before and they may come from different sectors. These are typically the collaborations of particle physics. Their participants take collaboration for granted.

The structure of collaborations is shaped by particular encumbrances or complexities imposed by these formative processes. Formal organizational structures are produced by persuasions from external sources, by uncertainties in the process of resource acquisition, by structural change in the resource providers, and by the involvement of parent organizations. These factors often propel scientists and managers into relationships with others whom they have few grounds for trusting. Formal procedures and a well-articulated division of labor well serve collaborations whose member organizations are suspicious of how the collaboration will affect their interests. Hierarchies and accountable decision makers work for collaborations whose participants were impersonally chosen and wish to remain as autonomous as possible. Professional managers are useful when instigators have gone to great lengths to convince patrons their plans were viable. In short, bureaucracy is an effective substitute for trust.

Collaborations must attend to their management in structured ways. Centralization and procedure can increase the sense of accountability in decision making. Alternatively, collaborations may limit their scope, increasing the autonomy of lower participants. Projects generated in conjunction with a reorganization of their funding source reorient the research interests of their members—the effect is to create multiple levels of authority. When disputes arise from competing organizational commitments, the mediation of middle management is available to resolve conflicts. In multi-sectoral collaborations such management—often provided by professional engineers—arbitrates between the divergent approaches of corporate and academic scientists.

Our findings suggest that when organizations *other* than universities are significantly involved in formation, participants should expect that professional engineering management supplied by corporate or governmental members will strongly influence collaboration structure. Such involvement is likely to help keep collaborations within budget. Collaborations that are technically aggressive can benefit from professional management, but these tend to be instigated by a single, dominant sector. When competition for resources in a specialty is keen, and a university-dominated collaboration proposes to develop novel instrumentation to distinguish itself from the rest of the field, central authorities at funding agencies or research facilities should consider pressuring the participants to include and empower engineers to deal with the managerial and budget problems that developing new instrumentation courts. Our results confirm the brokerage role of program managers both in the formation of collaborations and the resolution of conflicts between participating organizations.

In forming new collaborations, scientists and their administrators seeking precedents and models are well advised *not* to confine their search to collaborations in their particular specialty and agency. For example, drilling through various media has long been a geophysical research technique supported by the National Science Foundation. Yet the instigators of the second Greenland Ice Sheet Project, who wanted NSF in the 1980s to fund the drilling of a summit-to-bedrock ice core for geophysical analysis from the optimal locale on Greenland's glacier, would have found better guidance in the formation of Voyager (a 1970s NASA-funded space-science project) than in the formation of the Deep Sea Drilling Project (a 1960s NSF-funded project to obtain cores from the oceans' floors). Like Voyager, GISP2 took advantage of a stellar research opportunity that attracted so much interest that an impersonal, centralized competition for the selection of participants was needed. In contrast, the instigators of the Berkeley-Illinois-Maryland Array, though astronomers, would have benefited from knowing how the instigators of the Deep-Sea Drilling Project had worked through the difficulties of assigning primary responsibility to one organization for building a facility that all member organizations would share.

What can be learned from the classification of social objects—scientific collaborations, in this case? We tried and rejected many ordering principles during the course of the project.[4] Our fourfold structural classification is a beginning. Our experience is reflected by Geoffrey Bowker and Susan Leigh Star's description of a classification system as a forest in which "some areas will be left wild, or in darkness, or even unmapped" (1999: 32). The collaborations examined here are broader in range than previous studies, but they are still limited to the physical sciences and dominated by collaborations with American participants. It would be a mistake to view the analysis as identifying "natural kinds" or to view it as having causal efficacy. We hope our typology will push scholarly and policy debates beyond the tendency to view modern scientific collaborations as a matter of "one best way" and the equally pernicious view that there is nothing but overwhelming complexity. Classifications are generally evaluated according to whether they utilize consistent principles and whether their categories are mutually exclusive and jointly exhaustive. But classifications must also be useful. This usefulness, for us, depends on both the degree to which they *fit our historical understanding* of the projects and *relate to other dimensions of structure.*

We began with the dichotomy of craft and factory, a distinction that is appealing but empirically deficient on both sides. The factory image is based generally on studies done on Luis Alvarez's Berkeley group, in which, in the early 1960s, an assembly line of semi-skilled workers digitized hundreds of thousands of bubble-chamber photographs for further analysis by computer. The craftsperson image is a nostalgic appeal to a golden age set forth by some eminent physicists with jaundiced views on the multi-organizational collaborations that became essential for particle-physics experiments. It represents neither the state of affairs in collaborations generally nor the views of a wider selection of physicists. Our fourfold typology, though less romantic, is based on aspects of organizational variability that include the division of labor, leadership structure, formalization, and hierarchy.

Bureaucratic collaborations rely more heavily than other collaborations on a hierarchy of authority, on written rules and regulations, on formalization, and on a specialized division of labor. The importance of these attributes in scientific collaboration is not "efficiency" or "effectiveness" in any simple sense. Collaborations are characterized by voluntary ingress. Their ephemerality is an essential feature. Bureaucracy in scientific collaborations does not imply rigidity or impenetrable obtuseness, as it may where permanent organizations hire authoritarian personalities to manage indifferent subordinates and multiply rules to control every contingency. Bureaucracy in scientific collaborations protects organizational interests in collaborations where a small group of its members enter into agreements for resources, and in those that sharply distinguish science from engineering and vest each in particular organizations.

Leaderless collaborations do not have a single scientific leader but they are bureaucratic in other respects, with highly differentiated structures. We call them "semi-bureaucratic," because they rely on more collegial forms of management. A related "non-specialized" form is similar to the bureaucratic form in its hierarchical management structure but displays lower levels of formalization and differentiation. Only participatory collaborations, dominated by projects in particle physics, lack these Weberian features, which is why Knorr Cetina has described them as communitarian (1999). Particle physicists aim for consensual decision making. Their collaborations encompass an exceedingly broad range of operations, from decisions about the building of detector components, to cooperative examination of data streams, to questions of rights over specific analyses, appraisal of results, and the character of information provided to the public. This broad scope means that the collaboration as a whole limits the freedom of individual investigators. It is legitimate to wonder whether freedom is greater in a collaboration whose consensual governance extends to all aspects of creating knowledge, or greater in a collaboration whose teams operate with complete autonomy over a limited sphere.

The difference in kinds of freedom is the consequence, at the individual level, of our distinction between administrative and scientific formality. Informal administrative structures are sometimes associated with tight control over the development of instrumentation, data acquisition, data analysis, and even the external communication of results. In multiorganizational collaborations, the lack of bureaucracy in administration can imply relatively low autonomy in the general level of decisions one can take independently of one's peers. If the ultimate objective of scientific work is the presentation of new data, findings, and discoveries, then the most egalitarian collaborations are also the most tyrannical.

As was emphasized in chapter 3, bureaucracy is not a single dimension; collaborations may avoid some features of bureaucracy while employing others. Where research teams themselves wish to maintain significant independence in collecting data, formalization is necessary but multiple levels of authority may be undesirable. Aggregative projects that require independent teams to meet collaboration standards typically employ meticulous hierarchies to enforce these standards rigorously; or alternatively, they can, in effect, *end* the collaboration when information is collected and before analysis has started. The latter strategy has its own disadvantages, but is rendered feasible by the voluntaristic and ephemeral nature of collaborations. Where data analysis is collectively managed, less formalization is required. Limits on the freedom of teams to collect information as they might—should they possess sufficient resources of their own—can lead to substantial freedom in its manipulation and presentation.

Decisions regarding the construction of instrumentation will significantly affect the evolution, success, and outcomes of collaborations for as long as complex instruments are essential for science. One reason is directly related to resources. Mid-course changes in instrumentation design or subcontracting for materials or instruments are a common source of cost overruns. A second, argued in chapters 4 and 5, is that the desire to improve the quality of instrumentation competes with the less-than-popular constraints of schedule and budget. Participants in aggregative projects—those which pursue standardization of data from a variety of collection sites or multi-purpose instrumentation from a variety of contributors—are as likely as any others to want state-of-the-art instrumentation with customized bells and whistles, but they decrease their ability to collect this data if they cannot afford to equip all their sites. Moreover, innovation in one component of their instrumentation creates systemic demands on others. One common method of dealing with instrumentation is through subcontracting—the temporary addition of new organizational actors with strictly limited involvement. Their involvement in collaboration is positively correlated with bureaucratic management and cost overruns.

In the areas of greatest interest to management (time and money), the autonomy of structural elements remains critical. Projects exceed expected time frames when groups work independently to develop integrated instrumentation or to analyze shared data. The absence of enforcement mechanisms intrinsic to voluntary organizations means that research teams are at each other's mercy. Collaborations that avoided agreements to share data were more likely to be successful in the eyes of their participants. Once again, it bears emphasizing that unabridged collaboration has its demerits. A reduction of collaboration within collaboration may be conducive to a sense of control among individual members.

Collaboration is a form of social organization that inherently limits freedom. So long as one participates, one is bound by objectives, procedural agreements, collective understandings, and informal pressures. We have not said much about scientific performance. Even assuming the availability of adequate measures, the group of collaborations examined here is not well suited to the task. Our sample of collaborations did not include outright failures and it did not include projects that did not come to fruition at all. In any significant evaluation of project performance, such comparisons must be at the foreground of decisions about the selection of collaborations to study.

Interviews with participants did provide a means of addressing relationships between their subjective assessments of success and other experiences within the collaborative context. One particular focus was trust, given the recent surge of interest in this topic. But apart from the foundational trust enjoyed by scientific colleagues—indeed, among co-workers in all fields—we found little evidence that trust was generally important to scientific collaborations. This is not to say collaborations operate *without* trust—on the contrary, most of them appear very "trusting" most of the time, whether viewed from inside or out. But that is the case with many forms of social organization. Our evidence pertains to *variation* in trust: collaborations with more trust are not significantly different than those with less. In particular, they are not notably more successful in the eyes of their own participants than those with lower levels. The lone major exception to this generalization is that lower levels of trust in collaborations correlate with higher degrees of intra-collaboration conflict. The prominence of conflict in people's experiences and their distaste for it is the most likely explanation of the continuing belief in the importance of trust.

The interdependence of trust and conflict is much too dense to attribute causal or functional priority to either, but conflict is associated with bureaucracy, the necessary evil. Bureaucratic restrictions on freedom provide means of negotiating conflict. Less gratifying it may be, but collaborations can operate under conditions of conflict, and must often do so. As we have argued, collaborative work implies a reliance on the ability to divide work in ways that may not build trust among those without a history of trusting, and may generate distrust among those with it. In areas without the "communitarian" structures of particle physics, smaller work groups with autonomy in a delimited sphere provide the sense of freedom valued by most professionals. The non-collaborative realm exists within the collaborative. The generation and transmission of knowledge within single scientists and teams endures even within large collaborations.

The fundamental reason Max Weber, writing at the beginning of the twentieth century, saw bureaucracy as a superior organizational form and an important condition for economic modernization is that bureaucracies could make better use of *technical knowledge*. In the "rationalization" of modern life, science and bureaucratic organization were twin pillars, not polar opposites. It is ironic that in the twenty-first century, scapegoating is characteristic of the institutional relations between science and bureaucracy. The scientists we interviewed depicted bureaucracy as at best a necessary evil, at worst the source of all troubles. Why is there institutional antipathy between science and bureaucracy, between Truth and Efficiency?

The reason originates from the tradeoff between organizational coordination and individual freedom that is an inevitable feature of both bureaucracy and science. Control of activities by a collective implies and requires constraints on the actions of both individual scientists and individual organizations. Granted, the point of collaboration is to work *with* others and most collaboration is freely entered. But that is not all. The difference between cooperation and collaboration is that in the former, the interests of others are taken into account, while in the latter the interests of others may be overridden in the pursuit of the collective goal. Inter-organizational scientific collaboration entails constraints that do not allow scientists to go off on their own—not too far, anyway. If you have a better idea about what the collaboration should be doing, or about what you should be doing within the collaboration, you may not be allowed to do it. Science and bureaucracy are no longer properly characterized as allies, as in Weber's day, but best viewed as siblings. Because they are siblings, they have the right to fight.

Appendix A
Organizational Approaches to Collaboration

The Weberian concept of bureaucracy has been consistently useful in our examination of inter-organizational collaboration in science, but a vast body of research is grounded in other theories and techniques. Notably, a good deal of empirical work draws heavily on neo-institutional and network theories of organizations. Most of this literature does not specifically target scientific collaborations but instead shifts the attention to innovations, partnerships, alliances, and collaborative networks that span science, technology, and industry. There are two main lines of such inquiry. One strand of research stems from sociology and organizational theory and views collaboration as embedded in multiple relationships. The other is an economics approach that studies linkages, exchanges, and transactions flows in various inter-organizational collaborations (Powell 1998).

With the growing recognition that inter-organizational collaborations have grown dramatically in many scientific fields and industries, scholars have tried to establish generalizations that could be useful for the organization and management of the various forms of collaborations among a set of organizations, often crossing several sectors. Some authors realize that leading organizational actors face a dilemma—on the one hand, managers have a strong preference for autonomy and a healthy dose of discretionary power; on the other hand, they cannot dismiss the demand for innovation through collaboration that entails dependency on other organizations (Meeus, Oerlemans, and Hage 2001). Other researchers highlight the fact that collaborative relationships force firms to actively diversify because "heterogeneity and interdependence are greater spurs to collective action than homogeneity and discipline" (Powell 1998: 231). The lion's share of attention has gone to fields such as biotechnology, which are not only multi-disciplinary but also multi-institutional, including universities, private firms, research institutes, research hospitals, and government agencies.

One effort has been directed toward work on innovation in networks of organizations. Powell et al. (1996) studied a network of biotechnology firms, concluding that the locus of innovation is increasingly to be found in networks of learning rather than individual firms. It is chiefly through a "cycle of learning" that ties based in research and development enable organizations to access diverse sources of knowledge and to position themselves centrally in a competitive environment. Further empirical studies (e.g., Owen-Smith et al. 2002) demonstrate that innovative research in biomedicine is not only crucially dependent on networks of organizations, but the evolution and nature of these networks differ conspicuously between Europe and the United States. The American system of collaborative relationships between public research organizations, science-based biotech firms, and multi-national pharmaceutical corporations tends to be much more diversified and concentrated in dense regional clusters than the European system of collaborative university-industry relationships in biotechnology, which is sparser, with a smaller number of organizational partners, and mainly confined to national clusters.

Other literature on innovation through inter-firm collaboration draws heavily on theories of inter-organizational relations. However, it suffers from a proliferation of approaches and a multitude of uses of core concepts. Aldrich and Whetten (1981) tried to rescue the concept of inter-organizational networks from a purely metaphorical use by suggesting a theoretical distinction between three kinds of collective entities: organization sets, networks, and action sets. While all of these entities are, in a sense, "networks," they differ in their focus on specific organizational actors or time periods. An organization set is a specific group of organizations with which a focal organization is directly linked. An organizational network refers to a set of organizations linked by a certain kind of relationship, without regard to any particular organization. "Action sets" are groups of organizations that have forged a temporary alliance, such as the multi-institution scientific collaborations in our study.

Despite showing some promise (e.g. demonstrating the advantages of "linking-pin" organizations, which correspond to "bridges" in network terminology), this typology was not widely adopted. Although the depth and breadth of studies of inter-organizational relations are impressive, they have been of less use in our study of scientific collaborations than concepts from the traditional study of bureaucracy. Indeed several authors have noted that the multiplicity of studies have simply muddied the picture rather than making it clearer, compounding the fragmentation and heterogeneity of conceptual approaches and methods (Nohria and Eccles 1992; Alter and Hage 1993; Powell and Smith-Doerr 1994). An empirical examination of the references to various theories in interorganizational relations shows resource dependence theory is the leader, followed by social network theory, political power theory, and institutional theory (Oliver and Ebers 1998).

Another area of work is more narrowly focused on science and technology but not on university-industry linkages. This stream of research seeks to understand collaborative processes involving multiple organizations and to examine trends that underlie increased productivity. Many of these studies focus on collaborative publications as a reflection of integrative trends in modern science and technology. Noteworthy recent applications of this approach with relevance for science policy and research evaluation have included the assessment of interdisciplinary structures, the tracking of collaborative developments, and the link between scientific and technological human capital and collaborative relationships.

The first of these applications is exemplified by an attempt to overcome the largely field-specific biases of much bibliometric research and utilize a co-classification method (where co-occurrence of subject classification headings is used as a "mapping" principle) in order to quantitatively evaluate interdisciplinary clustering in energy research (Tijsen 1992). To triangulate, a follow-up mail survey was conducted with practicing scientists and science policy makers to elicit their opinion of the accuracy and appropriateness of interdisciplinary bibliographic mapping. Managers of research-and-development programs voiced skepticism and recommended that maps should not be used as a sole tool for research evaluation, setting of research priorities, allocation of funds, and management of programs, but be limited to use in conjunction with expert consultation and other assessment methods.

The second application is largely geared toward quantitative examination of scientific collaboration through co-authorship publication patterns over time (Leclerc et al. 1992; Luukkonen et al. 1992; Hicks and Katz 1996). In large part this body of research recognizes that, although collaboration has always characterized science and technology, the past several decades have witnessed a sharp increase in interdisciplinarity and collaboration (both national and international), which poses problems for science policy and calls for new methods of R&D evaluation (Hicks and Katz 1996). The overwhelming majority of the findings from constructing maps of co-authorship and co-citation patterns over time just confirm through sophisticated network means characterizations of changes in modern science like greater networking and collaboration, increased globalization, and growing interdisciplinarity (Gibbons et al. 1994; Ziman 1994). However, they rarely address such important issues as what these processes entail for the funding, planning, management, and evaluation of R&D; when they do (see, e.g., Hicks and Katz 1996), no concrete solutions are offered.

The third application probes into the dynamics of collaborative work involving multiple organizations and how homogeneity or diversity of human capital in multi-university research alliances affects publications productivity. Comparing two government-funded inter-university collaborations to push the limits of high-performance computing—one consisting only of astrophysicists who represented various universities and the other bringing together individuals from several universities and from different disciplines to advance high-end computer research on ecosystems—Porac et al. (2004) found that both extremes can be conducive to increased publication output but in somewhat different ways. While the monodisciplinary collaboration overall published a larger number of co-authored papers, the multidisciplinary, multi-university alliance had a higher rate of increase of co-authored articles between the period before the collaboration was formed and the period of active operation of the collaboration. Of course, it is difficult to generalize based on two cases, but this seems a promising avenue for future research.

A more economics-based approach is employed in a distinct line of research that uses network techniques to examine innovation through technology partnering, especially webs of strategic alliances. Often, samples of various sizes are quantitatively analyzed to discover industry-specific patterns of such alliances. Thus, a study of 1,700 strategic alliances (agreements with long-term goals where the transfer of new technology through R&D is central to the agreement) across 45 firms in information technology industries (computers, industrial automation, microelectronics, software, and telecommunications) at two five-year periods throughout the 1980s utilized various analytical tools (multi-dimensional scaling, cluster analysis) and network measures (density) to trace the evolution of such co-operation (Hagedoorn and Schakenraad 1992). The major finding was that network density of inter-firm collaborations had increased in all areas of information technology in the second half of the 1980s. However, when the pattern of alliances was compared to companies' market share there was only partial overlap, which indicates that network centrality does not necessarily translate into high ranking in the market hierarchy.

A similar analytical strategy for a comparable time span was adopted in an empirical assessment of the evolution of strategic alliances in several non-core industrial sectors (chemicals, aviation/defense, automotive, and heavy electronic equipment). Although density of strategic partnering uniformly increased in the second half of the 1980s, the dominance of the triad of the United States, Japan, and Europe has not resulted in greater internationalization of R&D spurred alliances (Hagedoorn 1995). As the author himself readily admits, the research is chiefly descriptive and empirically driven (ibid.: 225). To be sure, this is not necessarily a corollary of the use of network methodology, although the latter might be a contributing factor, since analysts have pointed out the virtual absence of a convincing theoretical framework for investigation of joint ventures or partnerships in general (Parkhe 1993).

The formal network apparatus has also been utilized in comparative cross-country evaluation of the innovative structure and performance of technological systems—a major interest of economists of innovation and of policy makers. Leoncini et al. (1996) adopted a holistic approach to address this issue in Germany and Italy. They constructed intersectoral innovation flow matrices by inputting three network measures: dispersion around the average value, density, and centralization. Overall, the findings demonstrate that, whereas the German technological system exhibits a fairly high level of systematic connections, in Italy the same system appears to be more segmented, with high-tech sectors co-existing alongside traditional, peripheral ones. For Germany the most significant channel of diffusion of innovation turns out to be manufacturing; for Italy it turns out to be the use of materials. One weakness of this otherwise thorough analysis is the apparent empiricism and data dredging. Despite these limitations, the potential of such highly sophisticated quantitative techniques for both short-term and long-term policy is substantial.

Appendix B
Selected Case Studies

Geophysics

Consortium for Continental Reflection Profiling
A collaboration of U.S. universities, formed in 1976, that adapted oil exploration techniques to probe greater distances into the crust. Funding from the National Science Foundation.

Deep Sea Drilling Project and Ocean Drilling Program
The Ocean Drilling Program and its predecessor, the Deep Sea Drilling Project, retrieve cores from the ocean floor using a dedicated research vessel that took advantage of techniques for offshore oil exploration. Funding from participating nations.

Greenland Ice Sheet Project 2 (GISP2)
GISP1, established in 1978, was a small collaboration of Danish, Swiss, and U.S. scientists, who established that analysis of ice cores from Greenland's glacier could yield a high-resolution climate record. GISP2 began in 1988 with more than 20 principal investigators from U.S. organizations, engineering management by the Polar Ice Coring Office, and science management by the University of New Hampshire. Funding from the National Science Foundation.

Incorporated Research Institutes for Seismology
A consortium of U.S. universities, established in 1984, to provide academic seismologists with digital instrumentation and the computation capabilities needed to work with digital outputs. Funding from the National Science Foundation.

International Satellite Cloud Climatology Project
An international collaboration, begun in 1982, to generate global statistics on the distribution and properties of clouds from weather satellites that provide daily regional coverage. Funding from member nations.

Parkfield Earthquake Prediction Experiment
A collaboration of government agencies and universities to concentrate instrumentation at a site that seismologists predicted would experience a major earthquake before 1993. Funding from the state of California and the United States Geological Survey.

Warm Core Rings
A multi-disciplinary collaboration, initiated in 1981, of oceanographic institutes and universities to investigate meso-scale eddies that separate from the Gulf Stream. Funding from the National Science Foundation.

World Ocean Circulation Experiment
An international collaboration to generate global data sets on ocean circulation. Field work began in 1990. Funding from participating nations.

Ground-Based Astronomy

Astrophysical Research Consortium
A collaboration of U.S. universities, formed in 1984, to build an innovative optical telescope and then to pursue a digital sky survey project. Funding from the member organizations, the National Science Foundation, and the Sloan Foundation.

Berkeley-Illinois-Maryland Array
A consortium that builds and operates a hard-wired array of antennae to create a high-resolution radio telescope at millimeter wavelengths. Collaboration formed in 1987. Funding from the member organizations and the National Science Foundation.

Hobby-Eberly Telescope
An international collaboration of universities to build an innovative optical telescope. Formal relations among the universities were established in 1992. Funding from the member organizations.

Keck Observatory
A collaboration principally of the University of California system and the California Institute of Technology, begun in 1984, to build and operate two innovative optical telescopes. Most funds come from the Keck Foundation through the California Institute of Technology.

Three Millimeter Very Long Baseline Interferometry
An international collaboration of radio observatories to demonstrate the feasibility of VLBI at wavelengths an order of magnitude lower than those previously used. Data were taken in 1981. Funding from member observatories.

Very Long Baseline Interferometry Consortium
A consortium of U.S. radio observatories, formed in 1975, to regularize their use for very-long-baseline interferometry at centimeter wavelengths.

Materials Science

Center for Polymer Interface and Macromolecular Assembly
A collaboration of U.S. universities and corporations, formed in 1994, to coordinate their research on polymer interfaces and macromolecular assemblies. Funding from the National Science Foundation.

Hybrid Organic/Inorganic Semiconductors
A collaboration of U.S. universities and corporations, formed in 1992, to study the semiconducting properties of various composites. Funded by Advanced Research Projects Administration of the Department of Defense.

Smart Materials Consortium
A collaboration of U.S. corporations and universities to develop a better vibration-canceling device. Funded by the Advanced Research Projects Administration, beginning in 1992, with costs shared by member corporations.

Science and Technology Center for Superconductivity
A collaboration of U.S. universities and a research institute to coordinate their research into high-temperature superconductivity. Funding from the National Science Foundation.

Advanced Light Source Beamline Collaboration
An international collaboration, formed in 1990, of universities and a corporation to build a beamline and end stations at the Advanced Light Source at Lawrence Berkeley Laboratory. Each organization raised its own funds.

Crystal Structure of CTA and CTP
A collaboration, formed in 1993, of a U.S. corporation, university, and research institute to determine the structure of a group of crystals with potential industrial applications. Funding from the corporation.

DuPont-Northwestern-Dow Collaborative Access Team
A collaboration of a U.S. university and two corporations to build and use a beamline and end stations at the Advanced Photon Source at Argonne National Laboratory. Each organization raised its own funds.

Positron Consortium and Participating Research Team
An international collaborations of universities, a U.S. research institute, plus a U.S. corporation to build and use a positron beamline and end stations. Funding from the National Science Foundation and the member organizations.

Medical Physics

Angiography Diagnostics
A collaboration of U.S. university medical centers, science department, and research institutes to develop a non-invasive technique for imaging coronary arteries. Multiple funding sources. Work began in 1979.

National Digital Mammography Development Group
An international collaboration, begun in 1991, of university medical centers and corporations to develop and test two systems for digital mammography. Funded by the National Institutes of Health, with costs shared by corporations.

Radiology Diagnostic Oncology Group
A series of collaborations of U.S. medical centers and a professional society to assess the efficacy of various radiological techniques for diagnosing various cancers. Funding from the National Institutes of Health.

Particle Physics

BNL-E-643: Experimental Studies of Antiprotonic Atoms in Gaseous H_2 and He and in Liquid H_2
A collaboration of U.S. universities that studied the magnetic moment of the antiproton, the sigma-hyperon, and kaon- and sigma-nucleus interactions in the 1970s.

BNL-E-650: Single Electron Production in Proton-Proton Collisions
A collaboration of U.S. universities to search, at Brookhaven's energies, for the surprisingly copious production of leptons found at other accelerators. Collaboration formed in 1974.

BNL-E-654: Search for Charmed Particles during Test Phase
A collaboration of U.S. universities and a research institute to test an innovative detector that each would subsequently use.

BNL-E-734: A Measurement of the Elastic Scattering of Neutrinos from Electrons and Protons
An international collaboration of universities and research institutes, formed in 1979, to test theoretically predicted results from these interactions.

BNL-E-791: Study of Very Rare K-Long Decays
A collaboration of U.S. universities and a research institute, formed in 1982, to search for processes whose existence would violate physical theory.

CLEO Collaboration at the Cornell Electron Storage Rings
A collaboration of universities that developed and used a general-purpose detector to take data at energies not covered by larger accelerator laboratories. Proposal approved in 1979.

FNAL-E-289: Small Angle Proton-Helium Elastic and Inelastic Scattering from 8 to 500 GeV
An international collaboration of universities and research institutes that performed experiments, in the 1970s, with innovative targets developed by Soviet physicists.

FNAL-E-398: A Further Study of Muon Nucleon Inelastic Scattering
An international collaboration of universities, formed in 1971, that developed a muon beamline to perform higher-energy versions of electron scattering experiments.

FNAL-E-616: Measurement of Neutrino Structure Functions
A collaboration of universities and a research institute that developed a narrow band neutrino beam to make precision measurements in the late 1970s.

FNAL-E-715: Precision Measure of the Decay Sigma to NE-NU
An international collaboration of universities and research institutes to investigate whether a particular particle decay violated accepted theory. Proposal approved in 1982.

SLAC-E-132: A Study of K-P Interactions Using LASS
An international collaboration of a research and universities to develop and use a detector specialized for the study of the strange quark. Proposal approved in 1977.

SLAC-E-137: Search for Low Mass, Metastable Neutral Particles
A collaboration of research institutes and a university to search for a hypothesized particle. Proposal approved in 1980.

SLAC-PEP-4-9: The Time Projection Chamber and 2-Gamma Detector
An international collaboration of universities and research institutes that developed the time projection chamber, which made significant advances in particle detection.

SLAC-PEP-6: The MAC Detector
An international collaboration of research institutes and universities that built a "magnetic calorimeter" that emphasized measuring particle energies over particle tracking. Proposal approved in 1977.

SLAP-SP-24: A Large Solid Angle Neutral Detector (The Crystal Ball)
A collaboration of U.S. universities and research institutes to develop a spherical detector, composed of sodium iodide crystals, for detecting neutral particles. Proposal approved in 1975.

SLAC-SP-32: MARK-III
A collaboration of U.S. universities and a research institute to develop and use a detector designed to collect data on known particles. Proposal was approved in 1981.

Space Science

Active Magnetospheric Particle Tracer Experiment
An international collaboration of research institutes, universities, and corporations that launched three spacecraft for independent and coordinated research into the Earth's magnetosphere. Proposal approved in 1978.

Einstein Observatory
A collaboration of U.S. research institutes, corporations, and universities to develop and use an orbiting x-ray telescope. Launched in 1978.

Giotto
An international collaboration of European research institutes and universities to build a spacecraft that flew by Halley's Comet. Launched in 1985.

International Sun-Earth Explorer
An international collaboration of research institutes, universities, and corporations to build a pair of satellites that could together resolve spacetime ambiguities in measurements of the magnetosphere and its boundaries. Launched in 1977.

International Ultraviolet Explorer
An international collaboration of research institutes to develop an orbiting ultraviolet telescope. Launched in 1978.

Voyager
A collaboration of research institutes and universities to develop a pair of spacecraft to take advantage of a rare planetary alignment that enabled the spacecraft to encounter multiple outer planets. Launched in 1977.

Computer-Mediated Collaborations

Center for Research in Parallel Computation
A recent collaboration of universities and research institutes to develop ways of making parallel computing as easy as supercomputing. Funded principally by the National Science Foundation

Grand Challenge Cosmology Consortium
A recent collaboration of universities and supercomputing centers to develop computation techniques for simulating cosmological processes. Funded by the National Science Foundation.

Upper Atmosphere Research Collaboratory
A recent international collaboration of research institutes and universities to develop software that enables scientists to coordinate their use of remote instruments. Funded by the National Science Foundation.

Notes

Introduction

1. Reported by Malcolm W. Browne of the *New York Times* ("Finances Worry Neutrino Researchers," June 6, 1998).

2. For the politics of "Big Science," the critical task is to characterize the conditions that can produce such a paradoxical event. Why is it that a stellar organization for producing knowledge cannot protect itself from taking a big budgetary hit when a rich nation has an economic downturn? The resolution of the paradox, we suspect, lies at the intersection of the project's genesis and its "multi-organizationality." Kamioka's science is "hot" or topical. Recent developments in cosmology have made many physicists and astronomers passionate about knowing whether the neutrino has a tiny mass or is truly massless, regardless of whether they have ever tried to use a water tank as a detector. Because the science is so topical, the prospect of success stimulates Nobel dreams. But topical science also gives a project an aura of impermanence. Participants in Kamioka need have no long-term commitment to developing the experimental technique of water-tank detectors and do not see the project as the foundation for decades of data, doctoral dissertations, or technological innovations. Most hope a decade from now to be working on whatever experimental technique is relevant to the next "hot" topic. The passion to know the mass of the neutrino suffices to make the individual participant work with the care, diligence, and discipline one associates with a great organization, but the participants cannot convincingly argue that the project is a foundation for the long-term health of permanent organizations.

3. Each of these topics is explored in the collection by Galison and Hevly (1992). Galison, in an attempt to articulate a single theme that links empirical studies, argues that the expansion of science on many fronts has forced scientists to confront the world outside their disciplines. This is undoubtedly correct, but it is equally true of small science. The "world outside" is important, but it is simply wrong to argue that it is a fundamental characteristic of "Big Science." Hevly, in his concluding summary, calls "Big Science" both "elusive" and "conveniently murky," recommending that scholars focus more sharply on the interaction of science and technology, institutional context, and the importance of collaborative research—precisely the themes we emphasize.

4. Throughout this work we use "multi-organizational collaboration" interchangeably with "multi-institutional collaboration" and "inter-organizational collaboration." We frequently use the shorter forms "collaboration" and "project" with the understanding that our study examines one particular type of structure.

5. Data should be broadly construed to include computer-generated simulations of natural processes as well as experimental or observational data.

6. That is why our penultimate chapter addresses technological practices and leads to our general conclusions. That is why we occasionally use the term "techno-scientific" to characterize the goals of collaborations.

7. We view an organization as a bounded identity in which some actors acquire rights to speak authoritatively for the whole, which is an essential difference between an organization and more diffuse social formations such as scientific specialties, research networks, and technical systems, all essential to modern science. This definition is close to that of Charles Tilly, but without his emphasis on categories (1998: 67), and draws on Harrison White's concept of identity. The collaborations examined here are themselves organizations, as are their constituents: universities, government laboratories, and private firms. However, collaborations typically have a stronger goal orientation and are designed to be ephemeral. Their purpose is transient (building a telescope, conducting an experiment) rather than continuous (educating students, generating new products).

8. In terms of absolute frequency, formal and informal collaborations between pairs of investigators, teams, and institutions are more common and have a longer tradition in the history of science than collaborations between three or more organizations. Our choice of three was determined by their essential "groupiness." Georg Simmel (1950) noted that groups of three actors differ from simple pairs by virtue of the fact that in groups of two, the departure of either party dissolves the group. Transposed to the present context, a scientific collaboration of at least three organizations does not necessarily (or even typically) terminate because one of the participants withdraws.

9. Our sample includes two particle-physics collaborations and one geophysics collaboration that involved Soviet institutes, but the analysis in the following chapters is based on a sample of projects in the U.S. Many of these have European affiliations, but we did not explicitly select collaborations for cross-national comparison. Note that many collaborations may involve multiple organizational sites without being inter-organizational in our sense. The coordination of industrial technology projects often involve multiple sites, and multiple R&D labs. For instance, Button and Sharrock (1998) describe three projects in one large firm that specialized in photocopiers. The largest project had more than 200 individual participants in its main development phase, including sites in the U.K. and the U.S. Another project employed more than 30 hardware engineers in the Netherlands and 18 software engineers in the U.K. Such projects have many similarities to multi-institutional collaborations—including problems of control and coordination, but they are subject to a single managerial hierarchy.

10. It is not uncommon for a set of research groups competing to solve common problems, to define their relationship as collaborative. Many formal and informal "networks" have this structure but this, too, involves a different sense of collaboration than the one employed here. For example, genetic scientists seeking to clone the myotonic dystrophy gene "collaborate" in a global research effort based on good will rather than formal agreement (Atkinson et al. 1998). Teams from Lawrence Livermore National Lab and MIT in the U.S., as well as Ottawa, Nijmegen, London, and Cardiff, met regularly over a period of several years to communicate findings and exchange resources such as reagents and probes. The "collaboration" consisted of these exchanges, while bringing together different approaches to the problem, based on different techniques. Members shared an objective, the cloning of the DM gene, and were funded by the Muscular Dystrophy Association, but differed in skills and access to various kinds of resources. Such projects are "collaborative" only in a weak sense. It is not clear that this variety of "collaboration" differs at all from the traditional model of scientific groups competing for priority in discovery (Merton 1973; Hagstrom 1964), described by Callon (1995) as the "competitive" model of scientific advance. The principal finding of Atkinson et al. is that the nature of this collaboration shifted over the course of time, as teams approached a solution to the problem, and divergent resources and approaches that provided for collaboration also provided "the basis for territoriality and competition" (1998: 268). Trust diminished, information was consciously and strategically withheld from other groups, and only those resources that were felt to be of limited benefit to the other teams were shared. Relationships that were already tense and competitive became especially rancorous between discovery and publication. As one junior scientist put it, these events dispelled "the myth of scientists working cooperatively and collaboratively together for the ulterior joint aim" (ibid.: 273). What occurred is nothing more than the conventional race for priority, with a common sponsor facilitating intergroup exchanges that readily foundered in the face of competitive pressures.

11. Quoted on page v of Galison 1997.

12. Hagstrom (1964: 251) described three factors necessitating modern collaborative arrangements: (1) the increasing expense of scientific facilities, (2) increasing complexity of research techniques and instruments, which require expertise that cannot be provided by a single individual, and (3) increasingly interdisciplinary research.

13. This distinction is lost when the concept "collectivization" is employed. For example, an increase in the number of authors per paper in leading astronomical journals (Fernandez 1998) by itself may indicate only an increase in the number of students and colleagues, or a change in normative practices allocating credit for research, rather than a shift in the organizational location of the research.

14. Knorr Cetina describes the historical roots of the growth in size of high-energy physics experiments. ATLAS, for example, involved 2,000 physicists and 200 physics institutes (1999: 159–166).

15. Ionospheric physics developed as a subfield of geophysics involving large-scale research of a collaborative nature that received particular impetus during the International Geophysical Year (1957–58). Astronomy, as a part of space science, had been a largely co-operative effort involving significant technological facilities for decades before the age of government largess (Tatarewicz 1990). Oceanographic researchers began to utilize large collaborations of multiple organizations in the 1960s and the 1970s, including five of the best-known multi-institutional projects—CLIMAP, MODE-1, GEOSECS, CEPEX, and MANGANESE MODULES (Mazur and Boyko 1981). The home page of the National Science Foundation includes in its list of disciplinary programs a link to "Crosscutting and Interdisciplinary Programs," that encourage scientists to draft multi-organizational proposals.

16. The database consists of approximately 4,800 journals in science and engineering fields, constituting a core set of international scientific journals.

17. Except where noted, these figures are from the National Science Board (1998).

18. Multi-authorship levels reach 70% or more in Italy, Portugal, Finland, Hungary, and Kenya.

19. Multi-authorship for all articles in physics increased from 28% in 1981 to 44% in 1995, while for U.S. physics it increased from 40% to 55%. For earth and space sciences, another focus of the present study, multi-authorship increased from 45% to 58% (National Science Board 1998). Luukkonen et al. (1992) report that physics, mathematics, and earth and space science are the fields with the highest levels of collaboration in the 1970s and the 1980s.

20. While bibliometric studies have their limitations, inferences from organizational affiliation can be relatively sophisticated. Kundra's (1996) investigation of Indian medical sciences in the first half of this century distinguishes between intra-departmental collaboration (within the same department), inter-departmental collaboration (between two departments within the same institution), inter-institutional collaboration (two organizations within the same country), and international collaboration (two organizations in different countries).

21. The bulk of scholarly attention has gone to the creation of infrastructure. Historical and sociological writing on particle physics is rich in accounts of accelerator laboratories: the politics of funding, the culture of management, and the pursuit of greater capacity (Hoddeson 1983, 1987, 1992; Needell 1983; Seidel 1983, 1986; Irvine and Martin 1985; Krige and Pestre 1986; Hermann et al. 1987, 1990; Traweek 1988; Heilbron and Seidel 1989; Krige 1989; Pestre 1989; Westfall 1989; Hayakawa and Low 1991; Wang 1995).

22. Methodological details are available in a series of reports available from the American Institute of Physics (1992, 1995, 1999).

23. The sampling procedure consisted of two stages: (1) selection of experiments to be studied and (2) selection of participants in those experiments to be interviewed (Guerrero 1993).

24. The selection of subjects was aided by consultations with the spokespersons for the project. Approximately 300 interviews on these experiments were conducted, transcribed, and analyzed. We note that this is not a random selection of experiments—indeed, there is a bias toward experiments considered successful in the eyes of the high-energy physics community. The final set of experiments analyzed here consisted of the following collaborations: BNL 643, BNL 650, BNL 654, BNL 734, BNL 791, FNAL 289, FNAL 398, FNAL 428, FNAL 616, FNAL 632, FNAL 715, SLAC-PEP-004-009, SLAC E-132, SLAC E-137, SLAC SP-024, SLAC-PEP-032, SLAC-PEP-006, SLAC-SP-007B, and one non-accelerator experiment—P-DECAY-IMB.

25. Space-science collaborations were selected to include major centers—the Goddard Space Flight Center, the Marshall Space Flight Center, the Jet Propulsion Laboratory, and the European Space Research and Technology Center—as well as various scientific interests and styles (e.g., planetary science and astrophysics; observatories, "fly-bys," and active experiments). Our space-science collaborations were the Active Magnetospheric Particle Tracer Experiment, the Einstein Observatory, Giotto, the International Sun Earth Explorer, the International Ultraviolet Explore, and Voyager. American participation in these projects was almost exclusively funded by NASA, European participation almost exclusively by ESA. The eight collaborations in geophysics and oceanography included the Consortium for Continental Reflection Profiling, the Deep Sea Drilling Project, the Greenland Ice Sheet Project, the Incorporated Research Institutes for Seismology, the International Satellite Cloud Climatology Project, the Parkfield Earthquake Prediction Experiment, the Warm Core Rings collaboration, and the World Ocean Circulation Experiment. Funding for the American part of these research efforts was provided mainly by the National Science Foundation, the National Oceanic and Atmospheric Administration, and the Office of Naval Research. International participation was funded by various sources (AIP 1995).

26. Seventy-eight interviews were conducted—our goal was to interview at least three participants in each collaboration. In contrast to the first two phases where different sets of questions were targeted to various types of participants and open-ended questions predominated, the last phase utilized a single, semi-structured questionnaire. The number of interviews ranged between two and six per collaborative project, with a mean of just over three. The interviewees were individuals in administrative positions or leadership roles in each of the main organizations involved in the collaboration.

27. For example, questions about the allocation of credit for individual publications proved relevant primarily for particle physics, where author lists typically include all members of the collaboration.

28. These consisted initially of eleven broad groups of factors with 90 subcategories. The collaborations in the first two phases of the project were also used for the analysis. Although the original interviews were relatively unstructured, this was possible since they were used as the basis for the identification of dimensions in the third phase. Where there were gaps in the data, we used telephone and email to clarify issues with our original informants.

29. An issue that required considerable thought was how to aggregate interview data from individuals to describe a collaboration as a whole. Our basic technique was as follows. First, we examined the information from each informant and averaged the values. Next, each aggregated variable was scrutinized in relation to the individual component scores to determine the most reasonable aggregate score. Aggregate file variables were recoded to reflect the closest approximation to this summary of the opinions of the individual informants who were involved in a particular collaboration. There is no way of eliminating judgment in this procedure, nor would a simple computational method be desirable. The mode, the median, or, in a few cases, the opinion of the scientific leader was used. The mode was used as the primary selection factor, since it represents the dominant opinion regarding the designated trait more accurately than the median and the mean. (In the case of a collaboration represented by three interviews, the mode would indicate an agreement by at least two informants.) However, in some cases there was no clearly defined mode, or there were two. Then the median served as a way to provide an estimate for a particular variable. Finally, if this last criterion also failed, the value produced by the interview with the scientific leader was generally used, but this step was rarely needed. Some of the most critical methodological issues raised in the context of a study such as this concern the sample of collaborations selected and the aggregation procedure. Truly, the devil is in these details. Disciplinary purists, whether historical or sociological, will reject the procedure as an unintelligible hybrid. From data transformation through aggregation such procedure depends on judgment as the foundation of aggregation: there is no *single* means of constructing multiple transcribed interviews as a collaboration-level measure. Which kind of summing ought to be employed? Should cases of extreme dissensus over the amount of conflict or trust be completely eliminated from the analysis? Is a technique appropriate for a collaboration with ten informants usable for a collaboration with two? Without immersion in the historical development of the project, such judgments are difficult, if not purely mechanistic. There is no shame in tradeoffs where tradeoffs are necessary.

30. Of course, since we were in the process of "discovering" the dimensions, many of the early interviews were incomplete sources.

31. Only 16 of the collaborations from the first phase are employed in the present analysis. We excluded a collaboration that conducted a non-accelerator experiment in an abandoned salt mine, one that used a rare emulsion detection technique, and one for which we were unable to obtain a sufficient number of interviews. A total of 53 collaborations are included in most of the analyses below.

32. The return rate on those follow-ups was about 60%.

33. To this end we employ quotations from our informants to illustrate our understanding of the processes occurring in these collaborations. This approach is related but slightly different from that of Collins (1998). It is generally possible to find quotations that illustrate *opposite* points, given a large number of interviews. What constrains us is our relationship-oriented methodology. BECAUSE a particular relationship is exhibited by our data, we select interview extracts that illustrate only *those* processes.

34. To a lesser extent, we used ISSCP, IRIS, Sagittarius A, and GISP-2.

35. In this volume we pay particular attention to nuclear and particle physics for two reasons. First, because this was the first branch of academic science to develop inter-organizational collaborations as a routine approach to the production of knowledge (Hevly 1992: 361). Second, because these areas of physics have been used as an important model of scientific and social organization (Knorr Cetina 1998).

36. A second pivot point is the method of typology, much misunderstood as a kind of forcing of reality into boxes. But even with the relatively small number of cases analyzed here, diversity is a problem as well as an opportunity. Modes of formation and organization are variable, but not infinite. What requires emphasis is that typologies are devices for simplification, but there need not be one typology that is best for all purposes. Significant variation among collaborations exists for all the dimensions in chapters 1–5, but the contribution of this study cannot be its mere demonstration—that has been accomplished by others. For instance, Maienschein (1993) proposed a threefold classification of collaborations in the biological sciences based on the reasons for collaborating: to promote an efficient division of labor, to enhance credibility, and to build community. But are the reasons for collaborating associated with their technological characteristics? Do collaborations based on certain kinds of technologies display consistent patterns of organization? More than one classification may be needed to adequately characterize the phenomenon.

37. Science and technology studies no longer employ the restricted, if still conventional, use of technology as instruments or hardware, but include the practices employed in design and utilization. For example, the technology of a medical-physics collaboration includes standardized procedures for data collection to produce compatibility in the analysis of treatments.

38. Indeed, one can say that collaborations *are* organization, but only in the trivial sense that all social formations are social organization. To speak of collaboration without organization is to speak of cooperation in our sense of taking the interests of others into account without some specified goal.

39. During a speech by the director of a $25 million science funding program, a physicist leaned over and said to his friend: "These damn bureaucrats go on interminably. Just give us a couple of slides and get off."

40. A local example serves to illustrate this point. During the course of this study we held a number of meetings with advisory committees, often consisting of those who were current or former heads of large, multi-institutional collaborations. In our final meeting we used the term "bureaucratic" to refer to one of the primary types of organization outlined in chapter 3, as the scientists reacted disapprovingly. In spite of repeated efforts to explain that bureaucracy is used in a descriptive and not an evaluative sense—all seemed to agree that the only possible meaning would be *negative*—it was nearly futile to get beyond the issue that some collaborations would be labeled "bureaucratic," which was felt to be unfair.

238   *Notes to Introduction and Chapter 1*

41. Time as well as space is part of the micro-macro dimension of social structure (Collins 1988).

42. It is important to note that this is not tautological: projects with formal organizational structures could, in principle, decide to take data collectively and share data extensively.

Chapter 1

1. Such a term is closely associated with a narrative approach to organizational analysis (Czarniawska 1997) and points to the importance of both the processes themselves and their telling, which can inform participants' understanding of their enterprise both during their project and after their collaborative has dissolved.

2. One astronomer vividly recalled the initial discussion over which direction a project would take: "[He] said 'Look, why don't we tell the dean we want to do both an optical and a radio project, but because of the timing, because we have to act more quickly, let's do the radio project first.' And to my astonishment, nobody argued against it, and everybody agreed with this, and I couldn't understand for the life of me why everyone didn't see that this would mean we would do the radio project and not do the optical project. But that was what happened. And after he said his thing, we immediately had a vote. I think there was one dissenting vote to this proposition, and that was that. That was the beginning."

3. For many years now, students of the research process have held that projects and programs of all varieties interact in dynamic ways with their environments. An exclusive focus on research scientists themselves can yield a distorted view of the development of science. Such terms as "technical systems" (Shrum 1985) and "techno-economic networks" (Callon 1985) point to the multiplicity of actors that employ and transform resources in the development of new knowledge (Gibbons et al. 1994).

4. At the extreme of the dichotomy of brokered and pre-existing relationships is a situation where the impetus for collaboration comes predominantly from the state sector. An example is one of the first large scientific collaborations in the developing world, the decision to build the synchrotron light national laboratory in Brazil (Velho and Pessoa 1998). Here the level of consensus within the wider community of physicists was relatively low, with support for the project stronger from branches of science outside physics. Arguments based on the prestige value of such a facility and the possibility that construction of such a machine would lead to the enhancement of local technological capabilities was greater among policymakers than from potential users within the Brazilian physics community itself. Even so the idea and impetus for the facility originated within a small group of individual scientists, cognizant of their need for political support.

5. The measurement of inter-personal context illustrates the methodological procedure that we have used throughout the study. Given the formative accounts that

typically appeared in our examination of high-energy physics, space science, and geophysics, we asked specifically about these factors in the third phase. Then, as described in the introduction, we returned to the collaborations in the earlier phases to code the same factors.

6. A "high" rating includes instances where collaborators had served together on advisory committees or review panels. Pre-existing relationships were coded as "low" when several of the major participants knew each other by reputation only. We used "medium" for collaborations built on relationships formed at conferences or other opportunities when participants enjoyed the opportunity to "talk shop" with their peers.

7. "Low" brokered relationships meant that the participants found each other through informal means. We used "medium" for collaborations whose ringleaders organized workshops to advertise their ambitions and to attract additional members.

8. For example, the Goddard Space Flight Center is a government organization whose scientists are public servants, whereas the Jet Propulsion Laboratory is administered by the California Institute of Technology under contract to NASA. Both design and build spacecraft for multi-organizational scientific missions.

9. We viewed the latter type as not having a dominant sector.

10. At the time of writing, "collaboration" has become such a positive icon that many organizations promote it *for its own sake*, rather than as a means to the end of accomplishing particular research goals. We will have more to say about this in the conclusion.

11. Cluster analysis is a standard statistical technique that arrays members of a set into groups by their similarities along pre-selected variables.

12. Four specific measures were used: pressure from parent organizations, the presence of a dominant sector, resource uncertainty, and pre-existing relationships among investigators.

13. Following social scientific convention, we use distance 10 as the cutoff point.

14. Common tactics of the instigators of such collaborations were to recruit their chief competitors in order to write an overwhelming proposal or to alert a resource-granting agency, hoping that the agency would provide help and guidance in finding the contributors that would make the proposal overwhelming.

15. APL, managed by Johns Hopkins University, was founded during World War II to develop proximity fuses and expanded after the war to embrace a variety of areas of relevance to the military, including the study of the space environment in which military satellites must operate. For background, see Dennis 1994.

16. The name of Schmerling's office and its position in NASA Headquarters changed in the course of agency reorganizations, but such developments likely had only a second-order effect on the development of AMPTE.

240   Notes to Chapter 1

17. AMPTE's goals are laid out and its instrumentation described in a set of papers in volume 23 of *IEEE Transactions on Geoscience and Remote Sensing* (1985).

18. Because MPIET was well along in development of another project involving the release of ions, AMPTE recapitulated much of what it was already doing, and the incremental cost of participating in AMPTE was sufficiently low that it did not need to raise more dedicated funds.

19. For a generic description of the need for IRIS and its scope, see Smith 1986.

20. Those interested in portable instrumentation held their workshop in Wisconsin in January, which seems an indication of their seriousness of purpose.

21. Not until 1965 did the NRAO complete its first telescope.

22. In fixed-target experiments at Fermilab, the accelerator smashed protons onto fixed targets, creating a hodge-podge of exotic subnuclear particles. A common trick of experimenters has been to filter out particular particles (e.g., hyperons) from the hodge-podge in order to observe how that particle decays or to set up an interaction between that particle and others.

23. In Winston's case, borrowed hardware (a lead glass array) eliminated the need to procure materials; finances would have been trickier if he had had to build his contribution "from scratch."

24. The results of the experiment contradicted Winston's earlier experiment and upheld the "standard model." See Hsueh et al. 1985.

25. Orbiting charged particles give off electromagnetic radiation in relation to the frequency of their orbit. This fact beats back efforts to increase the energy of accelerators for particle physics, but is a boon to all manner of scientists who can use the radiation as a probe to the structure of materials.

26. This is not to say it is a simple matter for researchers to interest each other in collaborating, turn their ambitions and speculations into concrete plans, and submit the plans for consideration by other organizational entities. But it is easier than expecting scientists researchers to work together without common interests, taking into account the politics of parent organizations, or seeking out patrons whose taste in proposals is unknown.

27. Throughout this volume, the relationships reported are significant at the 95% confidence level unless otherwise noted.

28. Recall that brokered collaborations are less likely to be based on pre-existing relationships. However, the dynamics of specific cases seem to indicate that it is brokerage that makes the difference.

29. This and all subsequent figures that display findings from cross-tabulations report only relationships that are significant at $p < .05$. Figure 1.2 is constructed from four bivariate cross-tabulations, each involving an association with brokered relationships: clusters of the four factors are shown for each level (low, medium,

and high) of brokering (presence of formal contracts, political attention, and large size (measured by many teams or organizations). For instance, the leftmost bars in each of the three clusters indicate the percentages for "presence of formal contracts" across the three values of brokering. We can read this relationship as follows: collaborations with medium or high brokerage at their inception exhibit a higher likelihood than those with low brokerage to have formal contracts (approximately 40% more collaborations from the first two groups as compared with collaborations with low extent of brokered relationships). Political attention (high/low) and size (many/few teams and organizations) are both coded as dichotomous. Size was collapsed into two categories from the original continuous variables: collaborations with fewer than seven organizations or teams may seem "large" in some absolute sense, but are "small" relative to the sample of projects examined here.

30. Indeed, Voyager started out larger than it ended up; NASA eliminated a selected team during construction because the team could not build its instrument within the weight limits to which it had agreed.

31. "Importance," in this sense, is in terms of the number of significant bivariate relationships with indicators of other dimensions. In many instances the presence of large numbers of particle-physics collaborations, which all had obvious sources of funding in DOE and NSF programs dedicated to supporting particle physics experiments, appeared responsible for the relationship. Consequently, we re-ran our analyses without particle-physics collaborations. In all the findings we report, the direction of the relationship is the same without their inclusion, though at reduced levels of statistical significance.

32. All the variables in the legend are dichotomous (yes = 1, no = 0) with the exception of decision-making style, which has three categories (hierarchical, more consensual than hierarchical, consensual). This figure compares the distribution of collaborations where scientific and administrative matters were settled through the 'chain of command' for projects with "certain" or "uncertain" sources of funding. The other two styles of decision making are excluded from the graph because "hierarchical decision making" allows the most clear-cut comparison. In short, where funding and other resources had to be "earned the hard way," a hierarchy was more often established than when collaborations utilized an established funding pattern (60% of the projects in the former category as compared with only 14% of the projects in the latter).

33. Each university handles its own public relations in BIMA, but in other cases, e.g. the Keck Observatory, a formal, professionally staffed public-relations office popularized the project and its accomplishments.

34. "Scientific representatives" are scientists who represent an organization or a specialty within a collaboration. Particle physicists often use the term "group leader;" space scientists and geophysicists often use "principal investigator."

35. Argonne could use NSF funds to hire postdocs, but not to pay the salaries of permanent staff members who performed Superconductivity Center research.

Had a corporate rather than a government-supported laboratory been a collaboration member, a formal contract would have been needed to deal with intellectual property issues.

36. In ground-based astronomy and particle physics, the university sector was almost always dominant in the instigation of collaborations.

37. Infrequent communication between teams is defined here as once a month or less often.

Chapter 2

1. The organizational characteristics of bureaucracy are treated here in relation to magnitude. The main discussion of bureaucracy is left to the following chapter.

2. There were other reasons: budgetary pressure; the political configuration in the legislative and executive branches of government; the end of the Cold War, which diminished the clout of particle physics as a stalwart of national security; the insistence of opponents that such enormous expenditures for a particle accelerator were probably unwarranted, given that appropriations for social programs were being cut. Last, but certainly not least, it was hard to make a case for expected practical applications, technological advances, and innovation spin-offs—exactly the salient advantages of the Human Genome Project, another monumental project in the same time period that was approved by Congress and eventually funded (Kevles 1997).

3. We note that although numbers, say, of organizations, may be fairly precise, perceptions of size are relative. Sayles and Chandler (1971) examined a multi-institutional effort of about 100 scientists and engineers who designed the Satan nuclear power reactor in the 1950s, which they characterized as a "relatively small size project."

4. Hall (1977) and Hage (1980) reached opposite conclusions on this question.

5. We were able to assess the level of resources indirectly using the relative expenditures on personnel and instrumentation (equipment) for about half of the sample but not for high-energy physics, space science, or geophysics. We asked respondents to estimate the percentage costs for personnel and those for equipment. In two-thirds of these collaborations, more than half of the funds were used to support personnel. In the most frequent case, more than two-thirds of the budget was spent on personnel, while less than one-third was spent on instruments, materials, and research facilities. At least for projects in these fields, numbers of participants serves as a reasonable proxy for the resources involved in a collaboration.

6. "Experiments think of themselves as 'collaborations' of autonomous units coming together 'freely' and on a relatively 'equal' basis to reach an 'understanding' about what they will contribute to the common goal." (Knorr Cetina 1998: 203) In this view, practices such as collective authorship, free circulation of

work between individuals, and responsibility for the working of the experiment as a whole are all counteracting tactics against individuation. But it is precisely for this reason that the analyst must independently assess the importance of people, teams, and organizations.

7. The number of research teams was counted in the same fashion way as the number of participants. We did not consider collaboration "teams" to include groups that performed specialized tasks for non-scientific as well as scientific projects. When collaborations enjoyed professional engineering or administrative services—e.g., the space-flight centers that build or contract for spacecraft for science projects—we counted one engineering team and did not inquire into subdivisions of their work.

8. The consequences include the relocation of the consensus process to the beginning of the experiment (Knorr Cetina 1995: 140).

9. This impression is confirmed by cluster analysis, which generates two major groupings, one that includes large and lengthy collaborations, the other including smaller and shorter collaborations.

10. In some instances, collaborations need little or no dedicated funding to pursue their research, but required permission to use facilities they did not control (e.g., astronomical observatories). The time it takes for such collaborations to obtain permission is a functional equivalent of the time it took most collaborations to obtain funding.

11. A lengthy period leading to funding can result in a project that is less costly and more efficient, as in the battle of laboratories (Cambridge and Stanford) from 1962 to the eventual support of the first U.S. electron-positron collider in 1970. SLAC was chosen, then submitted proposals every year incorporating various modifications. In the end, after re-labeling the project an experiment rather than a construction project in order to get it through Congress, it became less expensive and utilized superior technology than in earlier proposals (Paris 1998).

12. The recent dominance of colliding-beam over fixed-target experimentation in particle physics is eroding the flexibility of experimentalists. Accelerators that collide beams have a fixed number of places where the beams intersect, and experimentalists surround those points with concentrically nested detector components. There are no prospects for adding experimental sites to a colliding-beam accelerator, and improving or reconfiguring the components of a colliding-beam detector is far more technically daunting than for a fixed-target detector, whose components are linearly arrayed behind the target. Effective participation in a colliding-beam experiment requires collaboration at its inception or at the point of a major overhaul of its detector.

13. NASA Headquarters later urged a policy of "quicker, smaller, cheaper" construction of scientific spacecraft. The idea was that this policy would provide space scientists with some of the flexibility that laboratory scientists traditionally enjoy.

244   Notes to Chapter 2

14. This distinction has been institutionalized at NASA with separate grants programs to support research and development of instrumentation and contracts programs to support the construction and operation of spacecraft and their payloads of scientific instrumentation. A would-be contributor of a scientific instrument to a spacecraft uses grants to build and "space qualify" a prototype and then proposes a suitable variation on the prototype to fly on a spacecraft.

15. The preference of the leading scientists for expanding the collaboration's capabilities over keeping the collaboration intimate probably helped to alleviate resentments within Headquarters and among planetary physicists over AMPTE having been self-organized.

16. In the case of the U.S., there were jurisdictional conflicts between the Goddard project manager, who was legally responsible for the U.S. contribution, and the Applied Physics Laboratory engineer who directed the in-house construction of the spacecraft. The APL engineer performed most of the responsibilities traditionally associated with the project manager; the Goddard project manager's role was limited to international coordination and to managing NASA's direct contributions: tracking, data relay services, and launch vehicle services.

17. The teams were from MIT's Haystack Observatory, the University of California's Hat Creek Observatory, Caltech's Owens Valley Radio Observatory, the National Radio Observatory (Tucson), the Harvard-Smithsonian Center for Astrophysics (which operated the University of Massachusetts Observatory), and the National Radio Astronomy Observatory (Kitt Peak).

18. Interferometry is the technique of simultaneously observing the same object from different positions and combining the signals to generate an interference pattern, which can then be reconstructed into an image. The reconstructed image has superior angular resolution to the individual observations. The longer the baseline between observatories, the higher the angular resolution. The problem is that the signals received at distant observatories cannot be combined in real time but must be individually recorded. To obtain an interference pattern, the recordings must be made with exquisite precision and played back with exquisite synchronization.

19. The rare exceptions to this statement included a high-energy physics experiment in which a graduate student felt that a personal falling out between his advisor and another senior physicist in the experiment led to difficulties in publishing a paper based on his dissertation. In another high-energy experiment a senior physicist requested he be removed from the author list of papers he thought were not worth publishing.

20. The only place where students participated in one such collaboration was the development of software for remote observing. But that was an aspect in which the students had a self-interest (since they hoped to be remote observatory users), in which the professors needed inexpensive personnel for software development, and in which the collaboration's managers were comfortable with allowing scientists to revisit issues and try out new ideas. Thus the collaboration sought to seg-

regate bureaucratic aspects of the collaboration from the apprenticeship interactions that teach students how to behave as scientists.

21. These associations are evident when measured by the number of participants or the duration to funding. There were no significant correlations between decision making and magnitude when measured by numbers of organization, numbers of teams, or time to first publication.

22. Encroaching light pollution from expansion around San Jose was degrading observing conditions for users of the Lick Observatory.

23. The astronomer argued for a 10-meter monolithic mirror; the physicist argued for a 10-meter segmented mirror. The monolithic mirror was relatively straightforward to make, but its size made supporting and pointing it a huge technical challenge. The segmented mirror was easier to support and point (because its individual segments were lighter) but complicated to make because of the difficulties of making a symmetric hole out of segments.

24. Interviewees particularly cited Ed Stone, a planetary physicist who served on the Caltech faculty and in JPL management, as a particular powerful and effective member of the board.

Chapter 3

1. The term is from the French "bureau," meaning a desk with numerous compartments.

2. His full treatment, based on an analysis of nineteenth-century Prussian models, also included features such as the separation of the public and private lives of officials and administration as a full-time job—hardly necessary to mention in a modern context.

3. See, for example, Levin and White 1961, Van de Ven et al. 1974, Kuhn 1974, Pfeffer and Nowak 1976, Laumann and Pappi 1976, Koenig 1981, Zeitz 1985, Wiewel and Hunter 1985, and Alter and Hage 1993.

4. The situation is less uniform in Europe. Both individual nations and unions of nations build accelerator laboratories. In the latter case (e.g., CERN), the individual nations are still responsible for funding their physicists when they use the laboratory.

5. Particle physicists also perform experiments in mines, using the cosmic rays that constantly bombard the Earth as their "accelerator," and a few still use emulsions or bubble chambers as detectors at accelerators; however, these styles are very much the minority, and we did not include any in our sample.

6. In coding for scientific leadership, we followed French and Raven (1968), including persuasive as well as authoritative scientists as leaders: "Some people have better ideas, scientific ideas. . . . It typically doesn't happen that they boss other people around. . . . They try to convince the other person that this is really

a great experiment rather than telling the other person to do this experiment. That's the way leadership is exercised."

7. We include the case of a man—a scientist by training—who viewed himself as a project engineer.

8. The presence of an administrative leader is an aspect of "formalization" in the cluster analysis below, while the presence of a scientific leader is an independent dimension that differentiates two of the major type of collaboration.

9. When the authority of an administrative leader rested on a professional appointment that was independent of any scientist in the collaboration, we considered the administrative leader to be at least equal in rank to the scientific. NASA and ESA project managers are quintessential examples of administrative authorities with independent standing from any participating scientist in a space-science collaboration.

10. One aspect of leadership that is largely neglected in the present chapter is organizational. It is often easier to control a project if some of the difficulties of inter-organization coordination are made intra-organizational. The designation of "lead centers" and "host organizations" within some collaborations exemplifies such a structure. A "lead center" is an organization whose contribution in areas such as personnel made it "where the action was." Two-thirds of our collaborations had one (occasionally two) lead centers. A "host organization" provided essential services for the collaboration as a whole. Nearly all collaborations (94%), except those that required little or no dedicated funds, had host organizations—if for no other reason than to centralize responsibility for producing progress reports for funding agencies. The three collaborations that performed or supported very long baseline interferometry did not have host organizations. These collaborations supported themselves through funds and instrumentation their participants were already able to tap from radio observatories. The lead center of a collaboration was not always its host institution, as in the case of particle-physics collaborations where intellectual direction came from one of its university members and not from the physicists who worked for the accelerator laboratory. Such collaborations had a host institution (the accelerator laboratory where the collaborators all assembled to run the experiment) and a lead center, but the two were not the same.

11. Leaders that have played a large role in the formation of the collaboration may be the wrong choice for later stages. One participation characterized such a situation this way: "We got the strong signal from the community that X was the wrong guy to be our standard bearer. I mean, he is an interesting guy. He was very engaged with the project. He had a strong personal commitment to it. But he simply didn't have the people skills to sell it to the rest of the community. And in the rest of the community [our collaboration] got coded as 'X hoodwinked [the funding program] into giving him a million dollars a year.' And this was a huge part of . . . the resource pool. . . . So there was a lot of resentment against him."

12. Even when the project scientist in a space-science collaboration was also a principal investigator and had been instrumental in initiating the collaboration—

circumstances that led us to view the collaboration as having a scientific leader—the project scientist could choose not to exercise leadership: "I got telephone calls from the Chief Scientist [at agency headquarters]. He said, you really have to stimulate your guys to write more cross-collaborative papers, but it's my belief that you can't tell 50-year-old PhDs what they ought to be doing. The collaboration as a whole published something like 1,500 papers. Of those, very few were one-author papers, but they're not what the Chief Scientist used to call 'we the people' papers."

13. The two exceptions, both particle-physics collaborations, were exceptionally ambitious in pursuing new instrumentation that promised such vast improvements in detection capabilities that achieving functioning instrumentation was as much a goal as with making discoveries. In both cases, neither the scientific leader nor the administrative leader (who was a scientist consciously taking an engineering role) were willing to be disinterested bystanders to the other's concerns.

14. Outranking occurred in one-third of all collaborations in the data set. We did not find any cases in which a scientific leader owed his position to a participating engineer. Only one of our 53 collaborations (IRIS) had more than one engineering leader.

15. Since most of our interviewees had advanced degrees, university departments served as a common reference point in our interviews when discussing questions of hierarchy and authority. We asked our informants to use a typical university department in evaluating the degree to which the collaboration was hierarchically structured. That is, collaboration was viewed as "similar to a university department" when the collaboration was described as a unified organization consisting of a leader (or multiple leaders), autonomous researchers who agreed to be part of the collaboration, and those the autonomous researchers supervised. Interviewees in nearly 70% of the cases viewed their collaborations as similar to the structure of university departments or containing fewer levels of authority. In other collaborations, interviewees spoke of multiple lines of authority involving middle managers that leaders used to organize aspects of collaboration activities or of supervisory committees that oversaw the leaders and considered collaboration-wide issues or adjudicated intra-collaboration disputes. We coded such collaborations as having more hierarchy than a university department.

16. Formalization includes the presence of written contracts, an administrative leader, the division of authority, and the presence of internal and external formal evaluations. Hierarchy denotes the levels of authority, system of rules and regulations, style of decision making, and degree to which leadership subgroups made decisions. Strictly speaking, there is always some division of labor within the collaboration, so we used the codes "specialized" and "unspecialized" division of labor to distinguish between two distinct kinds of work division within the collaboration. The standardized four composite variables were subsequently submitted to cluster analysis, specifying a range of solutions from two to five clusters. The results from using squared Euclidean distance and Ward's method of clustering are presented in the dendrogram.

17. We sought an empirically derived classification since past research yields no sound theoretical grounds for postulating specific types of scientific collaborations. Exploratory factor analysis was used to test for common underlying concepts using principal components extraction. This provided an initial solution of four factors, and oblique rotation was used for a terminal solution. This group of dimensions yielded the highest Kaiser-Meyer-Olkin measure and the most significant chi-square for Bartlett's test of sphericity.

18. The groups are evident at rescaled distance level 5. Again the tradeoff is between a small number of groups (at the far right of the figure) with members that are more dissimilar and a larger number of groups (at the far left) that are highly cohesive. There is no magic formula for selecting, just as there is no magic number of groups in any particular analysis. Our criterion was interpretability: in view of the substantive details of the cases being clustered, which breakdown of collaborations makes sense?

19. The major alternative plan was for a telescope with a monolithic 10-meter mirror. The larger the mirror, the greater the telescope's magnification. The larger a monolithic mirror, the greater the mechanical difficulty of supporting the mirror and pointing and tracking with the telescope. The segmented mirror design reduced the mechanical problems while assuming the burden of developing controls that would keep the segments in a spherically symmetric whole.

20. This does not imply that trust is greater for collaborations formed from pre-existing relations, a question addressed in chapter 5.

21. We used DND-CAT as an example of an externally brokered collaboration in chapter 1.

22. Jerome Cohen, the Northwestern faculty member who most aggressively urged formation of the collaboration, had been a consultant to DuPont, but he resigned his consultancy to help manage the collaboration.

23. The chairman of the board was elected every year, rotating among Dupont (two years), Northwestern (two years), and Dow (one year), and the board was composed of representative in similar proportions. Initially, DuPont and Northwestern were to share costs equally, but with the addition of Dow to the collaboration Northwestern's and DuPont's shares were reduced to 40% each and Dow picked up the remaining 20%.

24. Unlike CARA personnel, these professionals were *de jure* employees of one of the participating organizations, since DND did not incorporate. This was purely an administrative convenience; it had no bearing on the substance of the work.

25. WCRP later sponsored the World Ocean Circulation Experiment (WOCE), which we also included in our database of collaborations, to address ocean-atmosphere interactions. Far more infrastructure had to be created for WOCE than for ISCCP. Thus ISCCP was viewed as a better first project for the new program.

26. In a similar manner, dissenters within Keck could appeal to the board of directors knowing that the board was inclined to support CARA.

27. Moreover, this type of organization for collaborations is the only one that overlaps clearly with one of the types—"collaborators wanted"—in terms of project formation. Both are dominated by particle-physics collaborations.

28. Few physicists, especially in the Eastern U.S., were involved in relativistic heavy-ion research and most interested scientists came from high-energy and nuclear physics.

29. The division of labor followed institutional lines: some focused on construction and assembly of apparatus, while others focused on data analysis and software development. During the construction phase, Brookhaven Lab served as the host institution and communications center, but this shifted to the State University of New York at Stony Brook as the project progressed to data analysis. Once the Brookhaven advisory committee expanded the scope of the project, all decisions were made within the collaboration. Teams presented progress reports at collaboration-wide meetings, which were used to provide teams the opportunity to petition for modification of an experimental run or apparatus in order to exploit a promising new area. Evaluations were dictated by Brookhaven's accelerator schedule. The scientific leader also chaired the Institutional Board, which consisted of one representative from each institution.

30. Although particle-physics collaborations dominate the participatory collaborations and tend to be more homogeneous in organization and management than other forms, this relationship, and those in chapter 4, are something more than a "particle-physics effect." Without particle-physics collaborations about half of all relationships fail to reach standard levels of statistical significance, but the direction of all association remains the same.

31. In figure 3.2, three cross-tabulations are combined into one bar chart. All aspects of communication and credit in the legend are binary (showing presence versus absence). For the bars we have only used the category that denotes presence of the respective attribute. The interpretation of the graph is straightforward. All participatory collaborations managed their external communication flows (submission of papers for publication or presentations at conferences), as compared with about 29% of either the leaderless or the non-specialized types. Only 12.5% of the bureaucratic collaborations did so. The other two variables—whether the collaboration witnesses a change in communication practices (typically, a somewhat sporadic communication during the project-formation phase was followed by intense interaction during the experimental run and a slowing down of this process in the data analysis phase) and whether the collaboration as a whole issued press releases—are interpreted in a similar fashion.

32. Rarely did a collaboration member sufficiently disapprove of a manuscript to request exclusion from the author list. In such cases, one informant told us, someone wishing to be excluded might say "'I just helped with the observations; I didn't do enough on the paper' . . . [in order] to preserve working relations [with other collaboration members]."

33. The relationships among faculty members within an academic department is one analogy.

34. The situation is more diverse in Europe. Both individual nations and unions of nations build accelerator laboratories. In the latter case (e.g., CERN), the individual nations are still responsible for funding physicists when they use the laboratory.

35. Particle physicists also perform experiments in mines, using the cosmic rays that constantly bombard the Earth as their "accelerator," and during our period of study a few still used emulsions or bubble chambers as detectors at accelerators; however, these styles are very much in the minority, and we did not include any in our sample.

Chapter 4

1. We treat accounts of project success in more detail in chapter 5.

2. Although data quality is always a concern of those who use data, it need not be an issue for the collaboration as a whole.

3. It is important to note that we did not select our collaborations for this characteristic, so it is significant that only ISCCP and RDOG (Radiological Diagnostic Oncology Group) had no instrumentation needs beyond what was freely available or available through their member organizations. GCCC (Grand Challenge Cosmology Consortium) and CRPC (Center for Research on Parallel Computation), which competed successfully for time on supercomputers, and Sagittarius A, 3mmVLBI, and VLBI Network, which competed successfully for time on radio observatories, were able to buy or borrow what the facilities they used did not provide.

4. The relationship between organizational type and subcontracting is discussed in the following section. In terms of subcontracting, about four out of eight leaderless and two-thirds of the bureaucratic collaborations resorted to this strategy, in contrast to only about one-third of the participatory and just one-sixth of the non-specialized collaborative projects in our sample.

5. Construction of instrumentation closely follows the design of instrumentation and generates significant differentiation among the three organizational types.

6. For example, the comparison of the rightmost bars under the "yes" and "no" codes for subcontracting can be interpreted as demonstrating that in half of the multi-institutional collaborations that subcontracted the building of instruments decisions were made almost exclusively by the leadership, whereas such decisions were made solely by the leadership in only 11% of collaborations that did not subcontract.

7. Since non-specialized forms typically do not design or construct equipment, no further explanation is needed.

8. Recall that "leaderless" collaborations have no single *scientific* leader.

9. The Voyager spacecraft had to be launched within a narrow period to take advantage of the planetary alignment that allowed navigation from one outer planet to the next.

10. Figure 4.3 shows that autonomy in data collection has a different relationship with organizational type than do agreements to share data. The sharpest contrast is exhibited by participatory collaborations—all projects of that kind worked out agreements to share data and none collect data separately by team. Participatory collaborations are, of course, the only field-specific type, dominated by high-energy physics projects.

11. Figure 4.4 plots five categories of acquiring data, ranging from "autonomous" to "integrated" dimensions of bureaucracy. These categories are significantly related to the dimensions listed in the legend. The less collective the collection of data, the more prominent are the Weberian features of bureaucracy (system of rules, clear-cut division of authority, and the presence of formal contracts). Just the opposite seems to be the case with decision making on scientific issues—while more than 92% of collaborations that took data in an integrated manner allowed scientific decisions to be made by the collaboration as a whole or the separate member organizations, only about one-third of collaborations in which the constituent teams collected data independently exhibited such a participatory style.

12. Overlaps in the scientific topics analyzed by structural groups largely parallels their agreements to share data, but the largest difference is between leaderless and non-specialized collaborations, with the former having significantly higher overlap. non-specialized collaborations are less likely to exhibit topical overlap—that is, scientific and social boundaries are more likely to correspond in these projects. For example, the groups in RDOG devised techniques to diagnose cancers of different organs (lung cancer, prostate cancer, colon cancer, pancreatic cancer).

13. Organizational interests are more closely related to the division of labor and participation in physics experiments at CERN owing to the national interests and resource relationships designed into the operation of the laboratory.

14. It is also typical that a single group takes the primary responsibility in an analysis and one person becomes its champion.

15. Compact disks were later produced for GISP2 data.

16. The question that remains in the relationship between data acquisition and formalization is how to interpret the category of "prescribed" data acquisition.

17. Results are the recorded and preserved traces of data collection; findings are knowledge claims based on these traces. We draw this distinction here because the "cross-checking" practiced between teams is closer to the original traces, though it is purposively related to the knowledge claims collaborators seek to generate. However, the distinction is often difficult to draw in practice, just as the difference between "raw" and "processed" data is shifting and contingent.

18. We will return to this point in the conclusion of the book.

19. This arrangement obliged the teams to make their data sets comprehensible to the scientific public. Some team members sought to view the obligation as an

opportunity more than a burden. One noted that "the data set that we deposited in the Space Science Data Center is generally much cleaner than it might have been.... I think the whole process [of archiving data sets] was generally positive, because it forced us to document and clean up the data in a way that we probably would have never done otherwise."

20. IRIS here stands for one of the major instruments on Voyager (Infrared Radiometer, Interferometer, and Spectrometer) and should not be confused with one of our other collaborations with the same acronym, the Incorporated Research Institutes for Seismology.

21. ISCCP further empowers its "customers" by making the merged data available in three stages of processing so that modelers can work with data that the global processing center has barely processed.

22. This device combines the features of a tape recorder and those of a computer.

23. In this respect they resemble non-specialized collaborations.

Chapter 5

1. Trust is clearly important, in a fundamental sense, to the constitution of social order and the construction of systems of knowledge. Indeed, the concept of "black-boxing" requires the idea of trust in certain kinds of knowledge claims and technological products. These well-known phenomena are not discussed further. The question we address is whether there is any special role of trust in large scientific collaborations.

2. Successful collaborations are said to require high inter-personal trust in "close and respectful" relationships among scientists, engineers, project managers, and administrators (Dodgson 1993). Trust enhances organizational effectiveness and long-lasting relationships between partners through technical, epistemic, social, administrative, and legal bonds (Hakansson and Johanson 1988).

3. In their framework for a research agenda in inter-organizational cooperation, Smith, Carroll, and Ashford claim that trust is "fundamental" and a particularly fruitful area for future research (1995: 15). See also Shapin 1994 and Kramer and Tyler 1995.

4. For two classic statements in the sociological literature, see Simmel 1950 and Coser 1956. Examples in the literature on inter-organizational relations include Assael 1969, DiStefano 1984, and Alter 1990.

5. This point is also standard in organizational theory (Pfeffer and Salancik 1978).

6. In science and technology studies (as in organizational studies, where multiple concepts—"effectiveness," "efficiency," "impact," "outcomes"—have emerged), the participant notions of "progress," "advance," and "development" and associ-

ated terms such as "discovery," "innovation," and "results" are used to express a positive evaluation of some research activity or event.

7. Classic examples are Pelz and Andrews 1966 and Irvine and Martin 1984. Productivity measures, to be applicable to collaborations, require some means for taking into account the idiosyncrasies of achieving results in different specialties. Efficiency measures require some means for taking into account the differences between collaborations that undertake significant instrumentation research and development and those that take advantage of extant instrumentation.

8. It would be less than truthful to suggest that we did not expect to find this relationship.

9. Our retrospective interviews do not allow us to draw firm conclusions about this.

10. For more than 20 years, trust has featured prominently in theories of social organization. Among the better-known treatments are Luhmann 1979, Giddens 1989, Fukuyama 1995, and Seligman 1997. Many distinctions have been drawn to refine the general notion of trust. Some authors, like Seligman, rely principally on the distinction between trust and confidence. In the literature of business and management, some distinguish among reliability, predictability, and fairness as components of trust (Zaheer, McEvily, and Perrone 1998). Others distinguish contractual trust (adherence to agreements), competence trust (competent role performance), and goodwill trust (commitment not to take unfair advantage) (Sako 1992).

11. Smelser (1997) illustrates this "sociological level" of trust with fiduciary roles, in which actors are legally mandated to act in terms of trust, irrespective of their psychological orientation toward each other.

12. Inter-personal and collective trust are not necessarily separate phenomena. Where only a single boundary relationship exists—say, between leaders (and no other members) of teams housed in different organizations—collective trust or distrust is an analytic distinction. Where multiple dyadic relationships exist between project teams, it is possible to have high collective trust even though distrust characterizes certain dyads, as well as the reverse situation, where high interpersonal trust does not correspond to high collective trust.

13. Projects with moderate levels of trust in other researchers were viewed as slightly more successful than projects with higher levels. No association with any of our performance measures was significant at the $p < .10$ level.

14. Paradoxically, this collaboration had the development of electronic media as one of its major objectives.

15. The same informant explicitly identified three strategies employed at various points in the project to resolve the conflict: direct, indirect, and coercive: "[There was] a kind of person-to-person strategy, so where . . . the X people have tried to identify someone who could 'get along with' the personnel at Y and try to understand them and sort of deal with them one-on-one either face-to-face or over the

phone. Then when that hasn't worked, we have tried to get influential intermediaries to try to identify to Y why it's in their best interest to go along with something. And then when that hasn't worked, we have simply fired shots across their bow by threatening not to renew their subcontract."

16. Note that the lowest category of trust toward project management has only two cases, while the lowest category of trust in other researchers has a single case. Excluding these categories does not change the significance of the results. Figure 5.1 is easier to interpret, since both conflict between teams and with management were dichotomized (presence versus absence) from the original ordinal variables. In their original form, the two "conflict" variables were coded as follows: 1 = not serious at all, 2 = not very serious, 3 = serious, 4 = very serious. The first two categories were collapsed into the "no" category; the last two were recoded as "yes."

17. Harry Collins has pointed out that, given the absence of a relationship between trust and performance, figure 5.1 is consistent with the notion that trusting collaborations are still "better" in the sense that collaborators prefer them.

18. The standard sociological wisdom, if such can be said to exist, is that conflict is positively associated with trust, but the reason is that multiple groups are involved. External conflict promotes internal solidarity.

19. We initially sought to examine a fourth axis of conflict: between junior and senior members of collaborations. However, since our informants were drawn from the senior members, we were unable to address this issue in a reliable way. One of our informants put it this way: "Since I was one of the few people in the beginning I had a lot to say about how it was done. I am not sure how happy I would be, involved as one of the soldiers." Indeed, high-energy physics collaborations often seem much less egalitarian to junior participants (Ulrike Felt, personal communication).

20. Recall that small collaborations were defined as having fewer than seven participating institutions, and large ones as having seven or more. Management of topics for analysis was a dichotomous variable. The interpretation of its [What is it?] association with conflict between project teams is straightforward: in more than half of the collaborations where there were serious disagreements among teams, analytical topics were managed by the collaboration as a whole (versus 27% of the collaborations in which there were no such conflicts).

21. Larger projects are also more formally organized, which may imply more established procedures for conflict resolution.

22. The collaboration irradiated sodium in the reactor to obtain its intense source of positrons. It was convenient for the collaboration to set up the beamline within the reactor facility, but moving outside the facility became worth the bother when the Department of Energy instituted stricter security over researchers working within the reactor.

23. Danish and Swiss scientists also participated in GISP-1. Other Europeans worked extensively on an Antarctic ice core extracted by the Soviet Union.

24. Hierarchy as a means of scientific organization is not inevitable simply because of the size of the collaborations involved. In many cases, either there is no scientific leader or there is a scientific leader whose discretionary power covers only administrative matters that are not scientifically meaningful. In many instances there is no external board to approve collaboration plans.

25. Consistent with this claim is a related finding that conflict between scientists and engineers is highest where there are frequent communications.

26. One participant was sufficiently relieved by the change that he wished the engineers' administrative authority had extended into software for data analysis, but the other physicists drew the line there: "There's a saying 'How much leadership can physicists tolerate?' People tend to be a bit too individualistic. They do not like to submit themselves to a common structure. . . . [We] needed a czar of data analysis and we really didn't have that. Gradually it got straightened out, because everybody wants the software to work, so you just keep patching it up. We didn't do nearly as good a job as we could have.'

27. Harry Collins' 1975 discussion of a "working" gravity-wave detector is the classic sociological delineation of this phenomenon.

28. Murnighan and Conlon (1991) examined the population of string quartets operating in Great Britain in the 1980s, interviewing nearly all of the participants.

29. A variety of methods have been developed to analyze performance in science, but no consensus exists on measures. For a review of methods based on counts, see Irvine and Martin 1985.

30. Two recent examples of this ambiguity are the loss of the Mars Orbiter in September 1999 and the failure of the Mars Polar Lander in early December of the same year. The Mars Climate Orbiter descended too rapidly and too far into the Martian atmosphere. According to one investigation, the trajectory was wrong because of a mathematical mismatch between the data that the Orbiter was sending to Earth (in the metric system) and NASA's reply from the ground (in the English system). Both Lockheed Martin, which manufactured the spacecraft, and JPL, which was in charge of the whole mission, admitted that negligence led to the loss of the Orbiter (London *Times*, November 12, 1999). The mishap with the Mars Polar Lander is more difficult to explain, although it is generally agreed that something disastrous happened during the descent. What merits attention in this case are not only the causes of these two particular accidents, but how success or failure is defined. Thus, the broader argument focused on the feasibility of NASA's new strategy, adopted in 1993, of "faster, cheaper, better." One could argue, for example, that the success of Mars Pathfinder and Mars Global Surveyor, in conjunction with the failure of the Climate Orbiter and the Polar Lander, was a reasonable return for $836 million dollars, especially in view of the total loss associated with earlier failures of "grand missions" (e.g. the Mars Observer).

31. The results of FNAL 715 may be written off as one more confirmation of the "standard model" for the structure of matter. Had the results contradicted the

standard model, they would have been touted as a point of departure for exploring a previously unrecognized facet of subnuclear particles. Even so, such judgments say nothing about the quality of the education FNAL 715 provided its graduate students or about the costs and benefits of close working relations between Soviet and American physicists in the 1980s.

32. Because many are academics and most are scientists, they are not as comfortable dealing with concepts (such as "effectiveness") that seem managerial in character.

33. For collaborations that build novel instrumentation advancing the state of the art, timeliness and cost effectiveness may seem trivial, since in the end participants are able to collect novel data. "Completion," moreover, can be defined flexibly: "I consider it highly likely that we will be largely complete by the declared completion date. I consider it even more likely that by fiat there will be a declaration of completion on that date. I feel 100 percent confident that the project will not be truly complete on that date."

34. Both measures refer back to the judgment of collaborators themselves. When we speak of external success, the reference is still to success *as described by our key informants*.

35. Since our measures were based on a selection of informants representing the scientists and the leadership, it is not surprising that most informants perceived their projects as very successful. After all, they had invested years of their professional careers, and sometimes they were responsible for millions of dollars in research monies. More noteworthy, perhaps, is that many informants, when comparing their collaboration against others they had experienced, qualified its successes in various ways. The most damning qualification is to focus on the performance of the collaboration "in context" (that is, its success in view of a project's many constraints), as in this example: "Well, given the limitations of funding and the slowness and development of machinery, I think everybody would agree it's been very highly successful." Of course, any collaboration may be considered successful, "all things taken into consideration."

36. One of the advantages of the methodology employed here is the reduced likelihood of inferring causality from correlation. For example, after examining collaborations that utilized both external advisory committees *and* experienced delays in completion, it seemed clear that the relationship between the two dimensions was an instance of a meaningful but non-causal relationship.

37. The central issue here is that when a project begins to exceed its time frame there often must be a coalition of support that includes both funding bodies and collaborating scientists. An external committee can marshal that support. In the private sector, the situation is parallel to continuous management support for a project. "It's just whether or not the project is seen as adding value to [private firm.] A lot of that goes with the management structure. You had management in place who valued this thing. . . . They'd go to their graves fighting for continued funding of this project. Well, when the people up there decide this is out of line,

the only way to get rid of [project] is to move those managers. When they went away, the people who were protecting the program disappeared. That was about it for the program."

38. In fairness, it should be noted that the European project re-used the drill for GISP-1, which had been built by a group spearheaded by the Danish, to drill all the way through the glacier to bedrock in order to retrieve the oldest samples of ice.

39. For example, no matter how clever the experimental design, if the accelerator did not produce new phenomena for detection, then no discoveries were possible. The converse was also true. Participants considered a collaboration successful because it had the good fortune of taking data at a fertile accelerator, even if the design of the detector was not especially clever. However, such cases were rare in our sample.

40. ARC was an entrepreneurial, observatory-building collaboration that failed to be "highly successful" because some interviewees emphasized that slippage in its schedule led not just to a delay in observatory operations but to poorer operations than anticipated: "I guess quite honestly I would have to say not very successful. . . . [It] look too long; it's still not working absolutely satisfactorily. I think it will eventually be a nice facility, but it's been operating now for a number of years and it's still got serious troubles."

41. The inverse relationship between size and conflict seems contrary to Schild's (1997) view that the size and the duration of polar research collaborations increase the potential for conflict. However, it may be precisely because of this potential that larger collaborations develop the means to control conflict.

42. Related support for this conclusion is found in Zaheer, McEvily, and Perrone's 1998 work on supplier relationships in the electrical equipment manufacturing industry. They show that complex trust is related to conflict, whereas inter-personal trust is *not* related to conflict and has no effect on performance. Furthermore, Dodgson, in a 1993 study of Celltech's relationship with the Medical Research Council, shows that difficulties with particular "opinionated" individuals, labor turnover, and even the termination of an important project manager, did not ultimately damage the inter-organizational trust between the two entities.

43. Recall the negative relationship between trust and bureaucracy, and the positive relationship between conflict and bureaucracy in collaborations. Organizations that initiate a collaboration trusting each other are much more likely to be able to form a collaboration with low levels of bureaucracy, while collaborations that can build trust quickly among participating organizations with no history of trust are more likely to avoid high levels of bureaucracy.

Conclusion

1. In our own collaboration, one point of continuing interest was the difficulty we had at times in distinguishing between certain elements of technology and

elements of organization. Indeed, it seems a simple matter when technology is only hardware—hence the continuing propensity to restrict its conceptualization. Yet when knowledge and utilization are viewed as crucial—as, indeed they must be—distinctions between organization and technology are considerably fuzzier.

2. These fields are readily identified by their participants. Two examples follow: "Collaborations are absolutely essential now because I don't think any of us have the breadth of skills necessary to carry out a project. You've got to know the astrophysics, you've got to know computing, you've got to know the hardware." "We're not discussing tabletop physics here. All progress in the field of relativistic heavy-ion physics is the result of collaborative efforts. They are physically enormous projects that require man-decades of work, and far beyond the scope of any individual."

3. This does seem to be the case in collaborations between developing and developed area, and is an underlying theme in the "collaboration paradox" described by Duque et al. (2005).

4. In the summer of 1997 we began work on an exploratory attempt to categorize the 23 collaborations in the last phase of this project, showing that of the major structural dimensions only "technological practice" was related to a broad range of outcome variables (Chompalov and Shrum 1999). That dimension, as in the present volume, did not focus exclusively on hardware, but incorporated the diverse ways that instrumentation and analytical tasks were organized to collect and process information. The analysis was only partly replicated on the broader sample of collaborations analyzed here, a sample that includes high-energy physics, space science, and geophysics. Though technology remains crucial in the structuring of scientific collaborations, its relationship with organizational factors is complex, as was indicated in chapters 3–5. Understanding the structure of scientific collaborations requires attention to both the patterns of behavior and artifacts that produce and process information.

References

Aldenderfer, M., and R. Blashfield, 1984. *Cluster Analysis*. Sage.

Aldrich, H., and J. Pfeffer. 1976. "Environments of Organizations." *Annual Review of Sociology* 2: 79–106.

Aldrich, H., and D. Whetten. 1981. "Organization-Sets, Action-Sets, and Networks: Making the Most of Simplicity." In *Handbook of Organizational Design*. Oxford University Press.

Alter, C. 1990. "An Exploratory Study of Conflict and Coordination in Interorganizational Service Delivery Systems." *Academy of Management Journal* 33: 478–501.

Alter, C., and J. Hage. 1993. *Organizations Working Together: Coordination in Interorganizational Networks*. Sage.

American Institute of Physics. 1992. *AIP Study of Multi-Institutional Collaborations. Phase I: High Energy Physics*. American Institute of Physics.

American Institute of Physics. 1995. *AIP Study of Multi-Institutional Collaborations. Phase II: Space Science and Geophysics*.

American Institute of Physics. 1999. *AIP Study of Multi-Institutional Collaborations. Phase III: Ground-Based Astronomy, Materials Science, Heavy-Ion Physics, Medical Physics, and Computer-Mediated Collaborations*.

Assael, G. 1969. "Constructive Role for Interorganizational Conflict." *Administrative Science Quarterly* 14: 573–581.

Atkinson, P., C. Batchelor, and E. Parsons. 1998. "Trajectories of Collaboration and Competition in a Medical Discovery." *Science, Technology, and Human Values* 23: 259–284.

Axelrod, R. 1984. *The Evolution of Cooperation*. Basic Books.

Bailetti, A., and J. Callahan. 1993. "The Coordination Structure of International Collaborative Technology Arrangements." *R&D Management* 23: 129–147.

Bailey, K. 1975. "Cluster Analysis." In *Sociological Methodology*, ed. D. Heise. Jossey-Bass.

Balmer, B. 1996. "Managing Mapping in the Human Genome Project." *Social Studies of Science* 26: 531–573.

Barton, A. 1955. "The Concept of Property-Space in Social Research." In *The Language of Social Research*, ed. P. Lazarsfeld and M. Rosenberg. Free Press.

Biagioli, M. 1998. "The Instability of Authorship: Credit and Responsibility in Contemporary Biomedicine." *FASEB Journal* 12: 3–16.

Bijker, W., T. Hughes, and T. Pinch, eds. 1989. *The Social Construction of Technological Systems: New Directions in the Sociology and History of Technology*. MIT Press.

Blau, P., and W. Scott. 1962. *Formal Organizations*. Chandler.

Blume, S. 1987. "The Theoretical Significance of Cooperative Research." In *The Social Direction of the Public Sciences*. 1987. Sociology of the Sciences Yearbook, Vol. XI. Reidel.

Bobrowski, P., and S. Bretschneider. 1994. "Internal and External Interorganizational Relationships and Their Impact on the Adoption of New Technology: An Exploratory Study." *Technological Forecasting and Social Change* 46: 197–211.

Boehme, G. 1997. "The Structures and Prospects of Knowledge Society." *Social Science Information* 36: 447–468.

Bowker, G., and S. Star. 1999. *Sorting Things Out: Classification and Its Consequences*. MIT Press.

Browning, L., J. Beyer, and J. Shetler. 1995. "Building Cooperation in a Competitive Industry: Sematech and the Semiconductor Industry." *Academy of Management Journal* 38: 113–151.

Bud, R., and S. Cozzens, eds. 1992. *Invisible Connections: Instruments, Institutions, and Science*. SPIE Optical Engineering Press.

Budros, A. 1993. "An Analysis of Organizational Birth Types: Organizational Start-Up and Entry in the Nineteenth-Century Life Insurance Industry." *Social Forces* 72: 199–221.

Button, G., and W. Sharrock. 1998. "The Organizational Accountability of Technological Work." *Social Studies of Science* 28: 73–102.

Callon, M. 1995. "Four Models for the Dyamics of Science." In *Handbook of Science and Technology Studies*, ed. S. Jasanoff et al. Sage.

Callon, M., J. Law, and A. Rip, eds. 1987. *Mapping the Dynamics of Science and Technology*. Macmillan.

Carper, W., and W. Snizek. 1980. "The Nature and Types of Organizational Taxonomies: An Overview." *Academy of Management Review* 5: 65–75.

Chompalov, I., and W. Shrum. 1999. "Institutional Collaboration in Science: A Typology of Technological Practice." *Science, Technology, and Human Values* 24: 338–372.

Clarke, A., and J. Fujimura. 1992. "What Tools? Which Jobs? Why Right?" In *The Right Tools for the Job*, ed. A. Clarke and J. Fujimura. Princeton University Press.

Clarke, L. 1989. *Acceptable Risk?* University of California Press.

Cohen, M., J. March, and J. Olsen. 1972."A Garbage Can Model of Organizational Choice." *Administrative Science Quarterly* 17: 1–25.

Cole, J., and S. Cole. 1973. *Social Stratification in Science*. University of Chicago Press.

Collins, H. 1974. "The TEA Set: Tacit Knowledge and Scientific Networks." *Science Studies* 4: 165–186.

Collins, H. 1975. "The Seven Sexes: A Study in the Sociology of a Phenomenon, or the Replication of Experiments in Physics." *Sociology* 9: 205–224.

Collins, H. 1998. "The Meaning of Data: Open and Closed Evidential Cultures in the Search for Gravitational Waves." *American Journal of Sociology* 104: 293–338.

Collins, R. 1998. *The Sociology of Philosophies: A Global Theory of Intellectual Change*. Belknap.

Coser, L. 1956. *The Functions of Social Conflict*. Free Press.

Czarniawska, B. 1997. *Narrating the Organization: Dramas of Institutional Identity*. University of Chicago Press.

Dennis, M. 1994. "'Our First Line of Defense': Two University Laboratories in the Postwar American State." *Isis* 85: 427–455.

DiStefano, T. 1984. "Interorganizational Conflict: A Review of an Emerging Field." *Human Relations* 37: 351–366.

Dodgson, M. 1993. "Learning, Trust, and Technological Collaborations." *Human Relations* 46: 77–94.

Durkheim, E. 1933 (1893). *The Division of Labor in Society*. Free Press.

Edge, D. 1977. "The Sociology of Innovation in Modern Astronomy." *Quarterly Journal of the Royal Astronomical Society* 18: 325–339.

Edge, D., and M. Mulkay. 1976. *Astronomy Transformed: The Emergence of Radio Astronomy in Britain*. Wiley.

Eisenhardt, K. 1989. "Building Theories from Case Study Research." *Academy of Management Review* 14: 532–550.

Encyclopedia of Sociology. 1981. DPG.

Etzioni, A. 1961. *A Comparative Analysis of Complex Organizations*. Free Press.

Etzkovitz, H., and L. Leydesdorff. 2000. "The dynamics of innovation: from national systems and 'Mode 2' to a triple helix of university-industry-government relations." *Research Policy* 29: 109–123.

Evan, W., and P. Olk. 1990. "R&D Consortia: A New U.S. Organizational Form." *Sloan Management Review* 31: 37–46.

Everitt, B. 1980. *Cluster Analysis*, second edition. Wiley.

Fairchild, H., 1977. *Dictionary of Sociology*. Littlefield, Adams.

Faulkner, R., and A. Anderson. 1987. "Short-Term Projects and Emergent Careers: Evidence from Hollywood." *American Journal of Sociology* 92: 879–909.

Fine, G. 1984. "Negotiated Orders and Organizational Cultures." *Annual Review of Sociology* 10: 239–262.

Freeman, C. 1991. "Networks of Innovators: A Synthesis." *Research Policy* 20: 499–514.

French, J., and B. Raven. 1968. "The Bases of Social Power." In *Group Dynamics*, third edition, ed. D. Cartwright and A. Zander. Harper & Row.

Friedberg, E. 1996. "The Relativization of Formal Organization." In *The Logic of Organizational Disorder*, ed. M. Warglien and M. Masuch. Walter de Gruyter.

Fukuyama, F. 1995. *Trust: The Social Virtues and the Creation of Prosperity*. Free Press.

Galison, P. 1987. *How Experiments End*. University of Chicago Press.

Galison, P. 1992. "Introduction." In *Big Science*, ed. P. Galison and B. Hevly. Stanford University Press.

Galison, P. 1997. *Image and Logic: A Material Culture of Microphysics*. University of Chicago Press.

Galison, P., and B. Hevly, eds. 1992. *Big Science: The Growth of Large Scale Research*. Stanford University Press.

Genuth, J. 1992. "Probe Report on the CLEO Experiment at CESR." In *AIP Study of Multi-Institutional Collaborations, Phase I: High Energy Physics*. American Institute of Physics.

Gibbons, M., C. Limoges, H. Nowotny, S. Schwartzman, P. Scott, and M. Trow. 1994. *The New Production of Knowledge: The Dynamics of Science and Research in Contemporary Societies*. Sage.

Gibson, D., and E. Rogers. 1994. *R&D Collaboration on Trial: The MCC Consortium*. Harvard Business School Press.

Giddens, A. 1989. *The Consequences of Modernity*. Stanford University Press.

Gilbert, N., and M. Mulkay. 1984. *Opening Pandora's Box: A Sociological Analysis of Scientific Discourse*. Cambridge University Press.

Gillespie, D., and D. Mileti. 1979. *Technostructures and Interorganizational Behavior*. Lexington Books.

Gillmor, C. 1986. "Federal Funding and Knowledge Growth in Ionospheric Physics, 1945–81." *Social Studies of Science* 16: 105–133.

Gray, B. 1985. "Conditions Facilitating Interorganizational Collaboration." *Human Relations* 39: 911–936.

Guerrero, A. 1993. Collaboration in Interorganizational Project Teams: A Study of High Energy Physics Teams. Ph.D. dissertation, University of California, Los Angeles.

Gulati, R. 1995. "Does Familiarity Breed Trust? The Implications of Repeated Ties for Contractual Choice in Alliances." *Academy of Management Journal* 38: 85–112.

Haas, J., R. Hall, and N. Johnson. 1966. "Toward an Empirically Derived Taxonomy of Organizations." In *Studies on Behavior in Organizations*, ed. R. Bowers. University of Georgia Press.

Hage, J. 1980. *Theories of Organizations*. Wiley.

Hagedoorn, J. 1995. "Strategic Technology Partnering during the 1980s: Trends, Networks, and Corporate Patterns in Non-Core Technologies." *Research Policy* 24: 207–231.

Hagedoorn, J., and J. Schakenraad. 1992. "Leading Companies and Networks of Strategic Alliances in Information Technologies." *Research Policy* 21: 163–190.

Hagstrom, W. 1964. "Traditional and Modern Forms of Scientific Teamwork." *Administrative Science Quarterly* 9: 241–264.

Hagstrom, W. 1965. *The Scientific Community*.

Hakansson, H., and J. Johanson. 1988. "Formal and Informal Cooperation Strategies in International Industrial Networks." In *Cooperative Strategies in International Business*, ed. F. Contractor and P. Lorange. Lexington Books.

Hall, R. 1977. *Organizations: Structure and Process*. Prentice-Hall.

Hatcher, L. 1994. *A Step-by-Step Approach to Using the SAS System for Factor Analysis and Structural Equation Modeling*. SAS Institute.

Hayakawa, S., and M. Low. 1991. "Science Policy and Politics in Post-war Japan: The Establishment of the KEK High Energy Physics Laboratory." *Annals of Science* 48: 207–229.

Heilbron, J., and R. Seidel. 1989. *Lawrence and His Laboratory: A History of the Lawrence Berkeley Laboratory I*. University of California Press.

Hermann, A., J. Krige, U. Mersits, and D. Pestre. 1987 and 1990. *History of CERN I and History of CERN II: Building and Running the Laboratory, 1954–1965*. North-Holland.

Hevly, B. 1992. "Reflections on Big Science and Big History." In *Big Science*, ed. P. Galison and B. Hevly. Stanford University Press.

Hicks, D., and J. Katz. 1996. "Where Is Science Going?" *Science, Technology, & Human Values* 21: 379–406.

Hickson, D., C. Hinings, C. Lee, R. Schneck, and J. Pennings. 1971. "A Strategic Contingencies Theory of Interorganizational Power." *Administrative Science Quarterly* 16: 216–229.

Hladik, K. 1988. "R&D and International Joint Ventures." In *Cooperative Strategies in International Business*, ed. F. Contractor and P. Lorange. Lexington Books.

Hoddeson, L. 1983. "Establishing KEK in Japan and Fermilab in the U.S.: Internationalism, Nationalism and High Energy Accelerators." *Social Studies of Science* 13: 1–48.

Hoddeson, L. 1987. "The First Large-Scale Application of Superconductivity: The Fermilab Energy Doubler, 1972–1983." *Historical Studies in the Physical and Biological Sciences* 18: 25–54.

Hoddeson, L. 1992. "Mission Change in the Large Laboratory: The Los Alamos Implosion Program, 1943–1945." In *Big Science*, ed. P. Galison and B. Hevly. Stanford University Press.

Hollingsworth, J. 1997. "Organizational, Network, and Institutional Influences on Major Discoveries in Bio-Medical Research in the United States." Presented at Conference on Linking Theory and Practice, Cologne.

Hsueh, S., et al. 1985. "Measurement of the Electron Asymmetry in the Beta Decay of Polarized Hyperons." *Physical Review Letters* 54: 2399–2402.

Irvine, J., and B. Martin. 1984. *Foresight in Science*. Pinter.

Irvine, J., and B. Martin. 1985. "Basic Research in the East and West: A Comparison of the Scientific Performance of High Energy Physics Accelerators." *Social Studies of Science* 15: 293–341.

Janoski, T., and A. Hicks, eds. 1994. *The Comparative Political Economy of the Welfare State*. Cambridge University Press.

Johnson, J. 1992. "Approaches to Organizational Communication Structure." *Journal of Business Research* 25: 99–113.

Kang, C., and R. Cnaan. 1995. "New Findings on Large Human Service Organization Boards of Trustees." *Administration and Social Work* 19: 17–45.

Kangas, O. 1994. "The Politics of Social Security: On Regressions, Qualitative Comparisons, and Cluster Analysis." In *The Comparative Political Economy of the Welfare State*, ed. T. Janoski and A. Hicks. Cambridge University Press.

Katz, D., and R. Kahn. 1966. *The Social Psychology of Organizations*. Wiley.

Keller, P., et al. 1982. "Measurement of the Electron Asymmetry in the Beta Decay of Polarized Hyperons." *Physical Review Letters* 48: 971–974.

Kevles, D. 1978. *The Physicists: A History of a Scientific Community in Modern America*. Knopf.

Kevles, D. 1997. "Big Science and Big Politics in the United States: Reflections on the Death of the SSC and the Life of the Human Genome Project." *Historical Studies in the Physical and Biological Sciences* 27: 269–298.

Kleinknecht, A., and J. Reijnen. 1992. "Why Do Firms Cooperate on R&D? An Empirical Study." *Research Policy* 21: 347–360.

Knorr Cetina, K. 1995. "How Superorganisms Change: Consensus Formation and the Social Ontology of High Energy Physics Experiments." *Social Studies of Science* 25: 119–149.

Knorr Cetina, K. 1999. *Epistemic Cultures: How the Sciences Make Knowledge*. Harvard University Press.

Knorr Cetina, K., and M. Mulkay. 1983. "Introduction: Emerging Principles in Social Studies of Science." In *Science Observed*, ed. K. Knorr Cetina and M. Mulkay. Sage.

Koenig, R. 1981. "The Interorganizational Network as a System: Toward a Conceptual Framework." In *The Functioning of Complex Organizations*, ed. G. England et al. Oelgeschlager, Gunn, and Hain.

Kramer, Roderick., and T. Tyler, eds. 1995. *Trust in Organizations*. Sage.

Kreiner, K., and M. Schultz. 1993. "Informal Collaborations in R&D: The Formation of Networks Across Organizations." *Organization Studies* 14: 189–209.

Krige, J. 1989. "The CERN Beam-Transport Programme in the Early 1960s." In *The Development of the Laboratory*, ed. F. James, Macmillan.

Krige, J. 1993. "Some Socio-Historical Aspects of Multi-National Collaborations in High Energy Physics at CERN between 1975 and 1985." In *Denationalizing Science*, ed. E. Crawford. Kluwer.

Krige, J. 1996. "The ppbar Project: II. The Organization of Experimental Work." In *History of CERN*, volume 3, ed. J. Krige. Elsevier.

Krige, J., and D. Pestre. 1986. "The Choice of CERN's First Large Bubble Chambers for the Proton Synchrotron (1957–1958)." *Historical Studies in the Physical and Biological Sciences* 16: 255–279.

Krige, J. 1998. "Constructing Credibility at/of CERN: The Boson Bonanza." Presented at meeting of History of Science Society, Kansas City.

Krimiges, S., et al. 1986. "AMPTE Lithium Tracer Releases in the Solar Wind: Observations Inside the Magnetosphere." *Journal of Geophysical Research* 91: 1339–1354.

Kuhn, T. 1970. *The Structure of Scientific Revolutions*, revised second edition. University of Chicago Press.

Kundra, R. 1996. "Investigation of Collaborative Research Trends in Indian Medical Sciences: 1900–1945." *Scientometrics* 36: 69–80.

Lammers, C., and D. Hickson. 1979. *Organizations Alike and Unlike*. Routledge and Kegan Paul.

Latour, B. 1987. *Science in Action: How to Follow Scientists and Engineers Through Society*. Harvard University Press.

Latour, B., and S. Woolgar. 1979. *Laboratory Life: The Social Construction of Scientific Facts*. Sage.

Laumann, E., Galaskiewicz, J., and P. Marsden. 1978. "Community Structure as Interorganizational Linkages." *Annual Review of Sociology* 4: 455–484.

Law, J. 1973. "The Development of Specialties in Science: The Case of X-Ray Crystallography." *Science Studies* 3: 275–303.

Law, J., ed. 1986. *Power, Action and Belief: A New Sociology of Knowledge*. Routledge and Kegan Paul.

Leclerc, M., Y. Okubo, L. Frigoletto, and J.-F. Miquel. 1992. "Scientific Co-operation between Canada and the European Community." *Science and Public Policy* 19: 15–24.

Lenski, G. 1994. "Societal Taxonomies: Mapping the Social Universe." *Annual Review of Sociology* 20: 1–26.

Leoncini, R., M. Maggioni, and S. Montresor. 1996. "Intersectoral Innovation Flows and National Technological Systems: Network Analysis for Comparing Italy and Germany." *Research Policy* 25: 415–430.

Lewis, J., and A. Weigert. 1985. "Trust as a Social Reality." *Social Forces* 63: 966–985.

Lewison, G., A. Fawcett-Jones, and C. Kessler. 1993. "Latin American Scientific Output 1986–91 and International Co-Authorship Patterns." *Scientometrics* 27: 317–336.

Lomax, R. 1992. *Statistical Concepts: A Second Course for Education and the Behavioral Sciences*. Longman.

Luhmann, N. 1979. *Trust and Power: Two Works*. Wiley.

Luukkonen, T., O. Persson, and G. Sivertsen. 1992. "Understanding Patterns of International Scientific Collaboration." *Science, Technology, and Human Values* 17: 101–126.

Luukkonen, T., R. Tijssen, O, Persson, and G. Sivertsen. 1993. "The Measurement of International Scientific Collaboration." *Scientometrics* 28: 15–36.

Lynch, M. 1985. *Art and Artifact in Laboratory Science: A Study of Shop Work and Shop Talk in a Research Laboratory*. Routledge and Kegan Paul.

Lynch, M., E. Livingston, and H. Garfinkel. 1983. "Temporal Order in Laboratory Work." In *Science Observed*, ed. K. Knorr Cetina and M. Mulkay. Sage.

Maienschein, J. 1993. "Why Collaborate?" *Journal of the History of Biology* 26: 167–183.

March, James G., and Robert I Sutton. 1997. "Organizational Performance as a Dependent Variable." *Organization Science* 8: 698–706.

Mazur, A., and E. Boyko. 1981. "Large-Scale Ocean Research Projects: What Makes Them Succeed or Fail?" *Social Studies of Science* 11: 425–449.

McKelvey, B. 1982. *Organizational Systematics: Taxonomy, Evolution, Classification*. University of California Press.

McMillan, G., F. Narin, and D. Deeds. 2000. "An Analysis of the Critical Role of Public Science in Innovation: The Case of Biotechnology." *Research Policy* 29: 1–8.

Meeus, M., L. Oerlemans, and J. Hage. 2001. "Sectoral Patterns of Interactive Learning: An Empirical Exploration of a Case in a Dutch Region." *Technology Analysis and Strategic Management* 13, no. 3: 407–431.

Merton, R. 1940. "Bureaucratic Structure and Personality." *Social Forces* 18: 560–568.

Merton, R. 1973. *The Sociology of Science*. University of Chicago Press.

Mezzich, J., and H. Solomon. 1980. *Taxonomy and Behavioral Science: Comparative Performance of Grouping Methods*. Academic Press.

Milner, M. 1994. *Status and Sacredness: A General Theory of Social Relations and Analysis of Indian Culture*. Oxford University Press.

Mitchell, G., ed. 1975 *A Dictionary of Sociology*. Aldine.

Mukerji, C. 1989. *A Fragile Power: Scientists and the State*. Princeton University Press.

Mullins, N. 1972. "The Development of a Scientific Specialty: The Phage Group and the Origins of Molecular Biology." *Minerva* 10: 51–82.

Murnighan, J., and D. Conlon. 1991. "The Dynamics of Intense Work Groups: A Study of British Quartets." *Administrative Science Quarterly* 36: 165–186.

Musselin, C. 1996. "Organized Anarchies: A Reconsideration of Research Strategies." In *The Logic of Organizational Disorder*, ed. M. Warglien and M. Masuch. Walter de Gruyter.

Narin, F., K. Hamilton, and D. Olivastro. 1997. "The Increasing Linkage between U.S. Technology and Public Science." *Research Policy* 26: 317–330.

National Science Board. 1998. *Science and Engineering Indicators—1998*. National Science Foundation.

Nebeker, F. 1994. "Strings of Experiments in High Energy Physics: The Upsilon Experiments." *Historical Studies in the Physical and Biological Sciences* 25: 137–164.

Needell, A. 1983. "Nuclear Reactors and the Founding of Brookhaven National Laboratory." *Historical Studies in the Physical Sciences* 14: 93–122.

Needell, A. 1987. "Lloyd Berkner, Merle Tuve, and the Federal Role in Radio Astronomy." *Osiris* 3: 261–288.

Newell, H. 1981. "The Broadening Field of Space Science." In *Space Science Comes of Age*, ed. P. Hanle and V. Chamberlain. Smithsonian Institution Press.

Nohria, N., and R. Eccles. 1992. *Networks and Organizations.* Harvard University Press.

Okubo, Y., J. Miquel, L. Frigoletto, and J. Dore. 1992. "Structure of International Collaboration in Science: Typology of Countries through Multivariate Techniques Using a Link Indicator." *Scientometrics* 25: 321–351.

Oliver, A., and M. Ebers. 1998. "Networking Network Studies: An Analysis of Conceptual Configurations in the Study of Interorganizational Relationships." *Organization Studies* 19: 549–583.

Orbuch, T. 1997. "People's Accounts Count: The Sociology of Accounts." *Annual Review of Sociology* 23: 455–478.

Orton, J., and K. Weick. 1990. "Loosely Coupled Systems: A Reconceptualization." *Academy of Management Review* 15: 203–223.

Owen-Smith, J., M. Riccaboni, F. Pammolli, and W. Powell. 2002. "A Comparison of U.S., and European University-Industry Relations in the Life Sciences." *Management Science* 48: 24–43.

Paris, E. 1998. "Lord of the Rings: SLAC, CEA, the AEC, and the Fight to Build the First U.S. Electron-Positron Collider." Presented at annual meeting of History of Science Society.

Parkhe, A. 1993. "'Messy' Research, Methodological Predispositions, and Theory Development in International Joint Ventures." *Academy of Management Review* 18: 227–268.

Parsons, T. 1960. *Structure and Process in Modern Society.* Free Press.

Pelz, D. 1963. "Relationships Between Measures of Scientific Performances and Other Variables." In *Scientific Creativity*, ed. C. Taylor and F. Burron. Wiley.

Pelz, D., and F. Andrews. 1966. *Scientists in Organizations: Productive Climates of Research and Development.* Wiley.

Perrow, C. 1967. "A Framework for the Comparative Analysis of Organizations." *American Sociological Review* 32: 194–209.

Perrow, C. 1972. *Complex Organizations: A Critical Survey.* Scott Foresman.

Perrow, C. 1984. *Normal Accidents: Living with High-Risk Technologies.* Basic Books.

Perry, N. 1993. "Scientific Communication, Innovation Networks and Organization Structures." *Journal of Management Studies* 30: 957–973.

Pestre, D. 1989. "Monsters and Colliders: The First Debate at CERN on Future Accelerators." In *The Development of the Laboratory*, ed. F. James. Macmillan.

Pfeffer, J. 1977. "Power and Resource Allocation in Organizations." In *New Directions in Organizational Behavior*, ed. B. Staw and G. Salancik. St. Clair.

Pfeffer, J. 1981. *Power in Organizations*. Pitman.

Pfeffer, J. 1982. *Organizations and Organization Theory*. Pitman.

Pfeffer, J., and G. Salancik. 1978. *The External Control of Organizations: A Resource Dependence Perspective*. Harper & Row.

Pickering, A. 1984a. *Constructing Quarks: A Sociological History of Particle Physics*. University of Chicago Press.

Pickering, A. 1984b. "Against Putting the Phenomena First: The Discovery of the Weak Neutral Current." *Studies in History and Philosophy of Science* 15: 85–117.

Pickering, A., ed. 1992. *Science as Practice and Culture*. University of Chicago Press.

Porac, J., J. Wade, H. Fischer, J. Brown, A. Kanfer, and G. Bowker. 2004. "Human Capital Heterogeneity, Collaborative Re4lationships, and Publication Patterns in a Multidisciplinary Scientific Alliance: A Comparative Case Study of Two Scientific Teams." *Research Policy* 33: 661–678.

Powell, W. 1998. "Learning from Collaboration: Knowledge and Networks in the Biotechnology and Pharmaceutical Industries." *California Management Review* 40, no. 3: 228–240.

Powell, W., K. Koput, and L. Smith-Doerr. 1996. "Interorganizational Collaboration and the Locus of Innovation: Networks of Learning in Biotechnology." *Administrative Science Quarterly* 41: 116–145.

Powell, W., and L. Smith-Doerr. 1994. "Networks and Economic Life." In *The Handbook of Economic Sociology*, ed. N. Smelser and R. Swedberg. Princeton University Press.

Price, D. 1963. *Little Science, Big Science*. Columbia University Press.

Price, D. 1984. "The Science/Technology Relationship, the Craft of Experimental Science, and Policy for the Improvement of High Technology Innovation." *Research Policy* 13: 3–20.

Pugh, D., D. Hickson, and C. Hinings. 1969. "An Empirical Taxonomy of Work Organizations." *Administrative Science Quarterly* 14: 115–126.

Ragin, C. 1994a. "Introduction to Qualitative Comparative Analysis." In *The Comparative Political Economy of the Welfare State*, ed. T. Janoski and A. Hicks. Cambridge University Press.

Ragin, C. 1994b. "A Qualitative Comparative Analysis of Pension Systems." In *The Comparative Political Economy of the Welfare State*, ed. T. Janoski and A. Hicks. Cambridge University Press.

Ragin, C., and Y. Bradshaw. 1991. "Statistical Analysis of Employment Discrimination: A Review and Critique." *Research in Social Stratification and Mobility* 10: 199–228.

Ragin, C., S. Mayer, and K. Drass. 1984. "Assessing Discrimination: A Boolean Approach." *American Sociological Review* 49: 221–234.

Ring, P., and A. Van de Ven. 1994. "Developmental Processes of Cooperative Interorganizational Relationships." *Academy of Management Review* 19: 90–118.

Rip, A. 1997. "A Cognitive Approach to the Relevance of Science." *Social Science Information* 36: 615–640.

Rip, A., and B. van der Meulen. 1996. "The Post-Modern Research System." *Science and Public Policy* 23: 343–352.

Roberts, E., and R. Mizouchi. 1989. "Inter-firm Technological Collaboration: The Case of Japanese Biotechnology." *International Journal of Technology Management* 4: 43–61.

Rogers, J., and B. Bozeman. 2001. "'Knowledge Value Alliances': An Alternative to the R&D Project Focus in Evaluation." *Science, Technology, and Human Values* 26: 23–55.

Sako, M. 1992. *Prices, Quality, and Trust: How Japanese and British Companies Manage Buyer-Supplier Relations*. Cambridge University Press.

Sanchez, J. 1993. "The Long and Thorny Way to an Organizational Taxonomy." *Organization Studies* 14: 73–92.

Samuel, Y., and B. Mannheim. 1970. "A Multidimensional Approach toward a Typology of Bureaucracy." *Administrative Science Quarterly* 15: 216–228.

Sayles, L., and M. Chandler. 1971. *Managing Large Systems*. Harper and Row.

Schild, I. 1997. "International Collaboration as Work in Polar Research." Paper presented at the annual meeting of Society for Social Studies of Science, Tucson.

Schott, T. 1991. "The World Scientific Community: Globality and Globalization." *Minerva* 29: 440–462.

Schott, T. 1993. "Science: Its Origin and the Globalization of Institutions and Participation." *Science, Technology, and Human Values* 18: 196–208.

Scott, J. 1991. *Social Network Analysis*. Sage.

Seidel, R. 1983. "Accelerating Science: The Postwar Transformation of the Lawrence Radiation Laboratory." *Historical Studies in the Physical and Biological Sciences* XIII: 375–400.

Seidel, R. 1986. "A Home for Big Science: The AEC's Laboratory System." *Historical Studies in the Physical and Biological Sciences* 16: 135–175.

Seligman, A. 1997. *The Problem of Trust.* Princeton University Press.

Shapin, S. 1994. *A Social History of Truth: Civility and Science in Seventeenth Century England.* University of Chicago Press.

Shapin, S. 1995. "Here and Everywhere: Sociology of Scientific Knowledge." *Annual Review of Sociology* 21: 289–321.

Shapiro, S. 1987. "The Social Control of Impersonal Trust." *American Journal of Sociology* 93: 623–658.

Shenhav, Y., W. Shrum, and S. Alon. 1994. "'Goodness' Concepts in the Study of Organization: A Longitudinal Survey of Four Leading Journals." *Organization Studies* 15: 745–766.

Shrum, W. 1985. *Organized Technology: Networks and Innovation in Technical Systems.* Purdue University Press.

Shrum, W. 1996. *Fringe and Fortune: The Role of Critics in High and Popular Art.* Princeton University Press.

Shrum, W., and J. Morris. 1990."Organizational Constructs for the Assembly of Technological Knowledge". In *Theories of Science in Society,* ed. S. Cozzens and T. Gieryn. Indiana University Press.

Simmel, Georg. 1950. *The Sociology of Georg Simmel.* Free Press.

Smelser, N. *Problematics of Sociology 1997.* University of California Press.

Smith, R. 1989. *The Space Telescope: A Study of NASA, Science, Technology, and Politics.* Cambridge University Press.

Smith, R. 1997. "Engines of Discovery: Scientific Instruments and the History of Astronomy and Planetary Science in the United States in the Twentieth Century." *Journal for the History of Astronomy* 28: 49–77.

Smith, K., S. Carroll, and S. Ashford. 1995. "Intra- and Interorganizational Cooperation: Toward a Research Agenda." *Academy of Management Journal* 38: 7–23.

Smith, S. 1986. "IRIS: A Program for the Next Decade." *Eos* 67: 213–219.

Sneath, P., and R. Sokal. 1973. *Numerical Taxonomy: The Principles and Practice of Numerical Classification.* Freeman.

Subramanyam, K. 1983. "Bibliometric Studies of Research Collaboration: A Review." *Journal of Information Science* 6: 33–38.

Tatarewicz, J. 1990. *Space Technology and Planetary Astronomy.* Indiana University Press.

Theodorson, G., and A. Theodorson. 1969. A Modern Dictionary of Sociology. Barnes & Noble.

Thompson, V. 1961. *Modern Organization*. Knopf.

Tilly, C. 1998. *Durable Inequality*. University of California Press.

Traweek, S. 1988. *Beamtimes and Lifetimes: The World of High Energy Physics*. Harvard University Press.

Traweek, S. 1992. "Big Science and Colonialist Discourse: Building High Energy Physics in Japan." In *Big Science*, ed. P. Galison and B. Hevly. Stanford University Press.

Trist, E. 1983. "Referent Organizations and the Development of Inter-Organizational Domains." *Human Relations* 36: 269–284.

Ullrich, R., and G. Wieland. 1980. *Organization Theory and Design*. Irwin.

Van Allen, J. 1983, *The Origins of Magnetospheric Physics*. Smithsonian Institution Press.

Velho, L., and O. Pessoa Jr. 1998. "The Decision-Making Process in the Construction of the Synchrotron Light National Laboratory in Brazil." *Social Studies of Science* 28: 195–220.

Wang, Z. 1995. "The Politics of Big Science in the Cold War: PSAC and the Funding of SLAC." *Historical Studies in the Physical and Biological Sciences* 25: 329–356.

Warglien, M., and M. Masuch, eds. 1996. *The Logic of Organizational Disorder*. Walter de Gruyter.

Warnow-Blewett, J. 1997. "Multi-Institutional Collaborations in Science: Documentation Issues." Paper presented at annual meeting of Society for Social Studies of Science, Tucson.

Weber, M. 1946. From *Max Weber: Essays in Sociology*, ed. H. Gerth and C. Mills. Oxford University Press.

Weinberg, A. 1961. "Impact of Large-Scale Science on the United States." *Science* 134: 161–164.

Weinberg, A. 1967. *Reflections on Big Science*. MIT Press.

Weingart, P. 1982. "The Scientific Power Elite—a Chimera: The De-Institutionalization and Politization of Science." In *Scientific Establishments and Hierarchies*, ed. N. Elias et al. Reidel.

Westfall, C. 1989. "Fermilab: Founding the First U.S. 'Truly National Laboratory.'" In *The Development of the Laboratory*, ed. F. James. Macmillan.

White, H. 1992. *Identity and Control: A Structural Theory of Social Action*. Princeton University Press.

Whitley, R. 1984. *The Intellectual and Social Organization of the Sciences.* Clarendon.

Wiewel, W., and A. Hunter. 1985. "The Interorganizational Network as a Resource: A Comparative Case Study on Organizational Genesis." *Administrative Science Quarterly* 30: 482–496.

Williams, A., E. Vayda, H. Stevenson, M. Burke, and K. Pierre. 1990. "A Typology of Medical Practice Organization in Canada." *Medical Care* 28: 995–1004.

Woolgar, S. 1988. *Science: The Very Idea.* Ellis Norwood.

Wuthnow, R. 1987. *Meaning and Moral Order: Explorations in Cultural Analysis.* University of California Press.

Wynne, B. 1989. "Sheep Farming after Chernobyl: A Case Study in Communicating Scientific Information." *Environment Magazine* 31, no. 2: 10–15, 33–39.

Zabusky, S. 1995. *Launching Europe: An Ethnography of European Cooperation in Space Science.* Princeton University Press.

Zaheer, A., B. McEvily, and V. Perrone. 1998. "Does Trust Matter? Exploring the Effects of Interorganizational and Interpersonal Trust on Performance." *Organization Science* 9: 141–159.

Zeitz, G. 1985. "Interorganizational Dialectics." *Administrative Science Quarterly* 25: 72–88.

Ziman, J. 1994. *Prometheus Bound: Science in a Dynamic Steady State.* Cambridge University Press.

Zucker, L. 1986. "Production of Trust: Institutional Sources of Economic Structure 1840 to 1920." *Research in Organizational Behavior* 8: 53–111.

Inside Technology: The Series

Janet Abbate, *Inventing the Internet*

Atsushi Akera, *Calculating a Natural World: Scientists, Engineers and Computers during the Rise of U.S. Cold War Research*

Charles Bazerman, *The Languages of Edison's Light*

Marc Berg, *Rationalizing Medical Work: Decision-Support Techniques and Medical Practices*

Wiebe E. Bijker, *Of Bicycles, Bakelites, and Bulbs: Toward a Theory of Sociotechnical Change*

Wiebe E. Bijker and John Law, editors, *Shaping Technology/Building Society: Studies in Sociotechnical Change*

Stuart S. Blume, *Insight and Industry: On the Dynamics of Technological Change in Medicine*

Pablo J. Boczkowski, *Digitizing the News: Innovation in Online Newspapers*

Geoffrey C. Bowker, *Memory Practices in the Sciences*

Geoffrey C. Bowker, *Science on the Run: Information Management and Industrial Geophysics at Schlumberger, 1920–1940*

Geoffrey C. Bowker and Susan Leigh Star, *Sorting Things Out: Classification and Its Consequences*

Louis L. Bucciarelli, *Designing Engineers*

H. M. Collins, *Artificial Experts: Social Knowledge and Intelligent Machines*

Paul N. Edwards, *The Closed World: Computers and the Politics of Discourse in Cold War America*

Herbert Gottweis, *Governing Molecules: The Discursive Politics of Genetic Engineering in Europe and the United States*

Kristen Haring, *Ham Radio's Technical Culture*

Gabrielle Hecht, *The Radiance of France: Nuclear Power and National Identity after World War II*

Kathryn Henderson, *On Line and On Paper: Visual Representations, Visual Culture, and Computer Graphics in Design Engineering*

Anique Hommels, *Unbuilding Cities: Obduracy in Urban Sociotechnical Change*

David Kaiser, editor, *Pedagogy and the Practice of Science: Historical and Contemporary Perspectives*

Peter Keating and Alberto Cambrosio, *Biomedical Platforms: Reproducing the Normal and the Pathological in Late-Twentieth-Century Medicine*

Eda Kranakis, *Constructing a Bridge: An Exploration of Engineering Culture, Design, and Research in Nineteenth-Century France and America*

Christophe Lécuyer, *Making Silicon Valley: Innovation and the Growth of High Tech, 1930–1970*

Pamela E. Mack, *Viewing the Earth: The Social Construction of the Landsat Satellite System*

Donald MacKenzie, *Inventing Accuracy: A Historical Sociology of Nuclear Missile Guidance*

Donald MacKenzie, *Knowing Machines: Essays on Technical Change*

Donald MacKenzie, *Mechanizing Proof: Computing, Risk, and Trust*

Donald MacKenzie, *An Engine, Not a Camera: How Financial Models Shape Markets*

Maggie Mort, *Building the Trident Network: A Study of the Enrollment of People, Knowledge, and Machines*

Nelly Oudshoorn and Trevor Pinch, editors, *How Users Matter: The Co-Construction of Users and Technology*

Shobita Parthasarathy, *Building Genetic Medicine: Breast Cancer, Technology, and the Comparative Politics of Health Care*

Paul Rosen, *Framing Production: Technology, Culture, and Change in the British Bicycle Industry*

Susanne K. Schmidt and Raymund Werle, *Coordinating Technology: Studies in the International Standardization of Telecommunications*

Wesley Shrum, Joel Genuth, and Ivan Chompalov, *Structures of Scientific Collaboration*

Charis Thompson, *Making Parents: The Ontological Choreography of Reproductive Technology*

Dominique Vinck, editor, *Everyday Engineering: An Ethnography of Design and Innovation*

Index

Accelerators, 13, 16, 32, 47, 68, 166
Active Magnetospheric Particle Tracer Explorer, 19, 35–37, 60–64, 73–77, 83, 84, 170, 186
Advanced Photon Source, 47, 101
Advisory Committee, 177–179
Applied Physics Laboratory, 36, 60, 61, 73, 83, 84
Argonne National Laboratory, 44, 47
Astronomy, 68, 201
  ground-based, 16, 32, 51, 197
  optical, 41
  radio, 41, 43, 54
Authority, 62, 74, 132
Autonomy, 119, 145–150, 171, 179, 193, 198, 209

Berkeley Illinois Maryland Association, 41, 54–56, 70, 185
Bibliometric studies, 7
Big Science, 1–3, 67, 68, 88, 207
Brookhaven National Laboratory, 14
Bubble chambers, 68
Budgets, 177, 184
Bureaucracy, 4, 20, 68, 87, 88, 175, 212, 213, 216

California Association for Research in Astronomy, 82, 99, 100
Case studies, 8–10
Centralization, 142, 210
CERN, 6, 14, 44, 87, 151
Cluster analysis, 20, 33, 89, 95, 96
Co-authorship, 7

Cohen, Jerome, 47, 48
Collaboration(s)
  bureaucratic, 22, 97–101, 128–130, 212
  externally brokered, 45–48
  formation of, 25–32, 208
  international, 7
  inter-organizational, 4, 167, 193, 202, 216
  large, 115, 117, 164, 190, 191, 208
  leaderless, 22, 97, 101–103, 128–130, 213
  multi-organizational, 3, 4, 20, 51, 89, 97, 119, 175, 191, 202
  non-specialized, 22, 97, 103–106, 128–130, 213
  organization of, 2
  participatory, 22, 97, 106–110, 128–130, 191, 213
  in particle physics, 87–89, 95, 108, 165, 166
  scientific, 2, 88, 90, 118
  technoscientific, 122
  theory of, 7–10
  typology of, 89, 95–110
  university-instigated, 60
Collins, Harry, 125
Communication, 6, 110–113, 204
  external, 106, 111
  management of, 55, 111
  patterns of, 110–113, 121
Communications center, 60, 61
Computer-centered collaborations, 17
Conflict, 18, 23, 152–154, 160–162, 215

278 Index

Conflict (cont.)
  between researchers and project management, 159, 162, 163, 167–171
  between scientists and engineers, 159, 163, 171–176
  between teams, 159–167
Consortia, 165, 166, 205
Contracts, 51, 74, 94, 107, 136
Cooperation, 2
Cornell Electron Storage Ring, 14
Credit, 110, 165, 188

Data
  acquisition of, 4, 53, 116, 120, 121, 125, 149, 150
  analysis of, 53, 113, 116, 120, 121, 125, 127, 214
  collection of, 122, 125–128, 135–141, 159
  management of, 40, 63, 64
  sets of, 38, 141
  sharing of, 53, 125, 126, 135–141, 214
Decision making, 74, 89, 116, 198, 210
  consensual, 94, 97, 117, 118, 193
  hierarchical, 94, 189, 210
  participatory, 190
Dendrograms, 95, 96
Department of Defense, 30, 37
Department of Energy, 29, 47, 58, 107, 184
Detectors, 52, 68, 106–109, 165, 166
Differentiation, 103
Distrust, 151, 152
Division of labor, 67, 87–89, 95, 97, 107, 181, 210
DuPont-Northwestern-Dow Collaborative Access Team (DND-CAT), 19, 47, 48, 101, 102, 130, 133, 137, 138, 143

European Space Agency, 9, 31
Experiments
  colliding-beam, 15
  fixed-target, 15, 43, 110
  hyperon, 43
  in particle physics, 69, 108, 139, 148, 189

Fermi National Accelerator Laboratory, 9, 14, 19, 43–45, 51–53, 56, 64, 80, 106–108, 113, 130, 134, 135, 139
Field sciences, 74, 167, 168
Formalization, 90, 95, 98, 103, 136, 141, 142, 212
Funding, 27, 72, 73
Funding agencies, 56, 179

Galison, Peter, 8, 11, 195, 196
Goddard Institute for Space Studies, 105, 106, 135
Governance, 116, 117
Greenland Ice Sheet Project, 140, 168–170, 178–180

Hagstrom, Warren, 6
Hierarchy, 87, 95, 97, 142, 212
History of science, 8
Hobby-Eberly Telescope, 73, 185

"Image" tradition, 11, 12
Incorporated Research Institutes for Seismology, 38, 50, 62–64, 70, 145
Instruments and instrumentation, 46, 59, 68, 69, 102, 119–123, 127, 128, 136, 198
Interdependence, 18, 119–121, 148–150, 171, 191, 193, 215
International Satellite Cloud Climatology Project, 103–106, 129, 130, 141, 145, 146
International Ultraviolet Explorer, 72, 73, 170, 182
Inter-personal context, 27, 28

Jet Propulsion Laboratory, 46, 82, 100, 134

Kamioka Neutrino Observatory, 1–3
Keck Observatory, 19, 81, 82, 99, 100, 130, 133, 174, 185
Knorr Cetina, Karin, 3, 25, 69, 70, 88, 115, 118, 153, 154
Kuhn, Thomas, 154

Laboratories
  accelerator, 115, 118, 139
  corporate, 57
  government, 57
  national, 68
Laboratory sciences, 74, 167, 168
Lach, Joseph, 43, 44
Leaders
  administrative, 74, 91, 132
  scientific, 90, 91, 95, 97, 107
Lederman, Leon, 9
"Logic" tradition, 1–12

Magnitude, 20, 67–74, 207
  and decision making, 79–85
  and formalization, 74–78
Materials research, 17, 57, 68, 201
Matthew Effect, 6, 11
Max Planck Institute for Extra-Terrestrial Physics, 36, 60, 76, 83, 84
Memorandum of understanding, 44–46, 106, 107
Merton, Robert, 6, 125

National Academy of Sciences, 46
National Aeronautics and Space Administration, 9, 31, 36–39, 46, 47, 75, 76, 83–85, 133, 134, 182, 183
National Oceanic and Atmospheric Administration, 146
National Science Foundation, 29, 54, 58, 107, 184, 211
Nobel Prize, 151

Observatories, 32, 41, 68, 112, 147
Oceanography, 15, 68
Organization
  formal, 132
  hierarchical, 55
  and management, 90
Organizational context, 31, 62–64

Particle physics exceptionalism, 97, 110, 115, 150
Performance, 151–155, 174–177, 214, 215

Physics
  geo-, 15, 16, 32, 50, 90, 200, 201
  heavy-ion, 108, 109
  high-energy, 12, 15, 88, 139, 183, 188
  medical, 17, 50, 188, 189
  particle, 6, 14, 32, 43, 51, 53, 68–70, 88, 139, 175, 194, 199, 206, 209 (*see also* Particle physics exceptionalism)
Political imperative, 198, 199
Positron Diffraction and Microscopy Collaboration, 165
Press releases, 111–113
Principal Investigators, 93, 139, 144, 171
Project formation, 25–32, 64–66
Project management, 158, 167, 171, 182, 197
Publication, 73

Relationships
  brokered, 27, 28, 192
  pre-existing, 27, 28, 45, 107, 109, 157, 159, 192, 210
  social, 204
Resource uncertainty, 27–29, 51–56, 183, 192, 208
Results
  cross-checking of, 120, 127, 142, 146, 147
  dissemination of, 114, 116
Rubbia, Carlo, 9, 151
Rules
  system of, 62, 74, 94
  written, 87, 212
Rutherford-Appleton Laboratory, 60, 76

Science and Technology Center for High-Temperature Superconductivity, 58
Science and technology studies, 11
Science Management Office, 16, 140
Science Steering Committee, 82, 100, 112
Sectoral context, 27, 30, 31
Sectoral instigation, 60–62

Sectors, 33–40, 45, 60, 201, 209
Seismology, 38, 39, 50, 62, 63
Social studies of science, 87
Sociology of science, 10
Space science, 15, 32, 50, 89, 200
Stanford Linear Accelerator Center,
    4–9, 19, 32, 149, 165, 166, 172, 173
Subcontracting, 122, 130–135, 180,
    182, 214
Success, 70, 176, 177, 183–189
  external, 187–189
  internal, 187, 189
  perceptions of, 158, 186
Superconductivity, 58, 59

Teams and teamwork, 4, 5, 60, 61, 71,
    120, 158, 159, 186, 197, 207
Technological determinism, 197
Technological innovation, 127, 195,
    196
Technological practices, 20, 120, 121,
    127, 198, 199, 202, 203
Technology, 3, 23, 119–121, 149, 173,
    196–198, 203
Technoscience, 20
Topics, management of, 126, 142,
    148–150, 193
Trust, 3, 23, 151–153, 162, 163, 191,
    192, 215
  collective, 157
  foundational, 157, 159

Van Allen, James, 35
Very long baseline interferometry, 73,
    77, 78, 109, 114, 147, 201, 202
Vorobyov, Alexei, 44
Voyager, 19, 46, 47, 114, 133–135, 140,
    143

Weather prediction, 103
Weather satellites, 103, 104
Weber, Max, 87, 97, 215, 216
Winston, Roland, 44
World Meteorological Organization,
    103

RETURN TO: PHYSICS-ASTRONOMY LIBRARY